Sustainable Approaches to Urban Transport

Sustainable Approaches to Urban Transport

Edited by
Dinesh Mohan and Geetam Tiwari

CRC Press
Taylor & Francis Group
Boca Raton London New York

CRC Press is an imprint of the
Taylor & Francis Group, an **Informa** business

CRC Press
Taylor & Francis Group
6000 Broken Sound Parkway NW, Suite 300
Boca Raton, FL 33487-2742

First issued in paperback 2021

ISBN 13: 978-1-03-209176-1 (pbk)
ISBN 13: 978-1-138-54423-9 (hbk)

Visit the Taylor & Francis website at
http://www.taylorandfrancis.com

and the CRC Press website at
http://www.crcpress.com

Publisher's Note
The publisher has gone to great lengths to ensure the quality of this reprint but points out that some imperfections in the original copies may be apparent.

Contents

Preface

More than a century ago, Ebenezer Howard created a movement centred on satellite cities enabled by rail transit access, focused primarily on real estate development with rail as the primary conduit between developed areas (Carlton, 2007). About a century later, Peter Calthorpe codified the concept of transport-oriented development (TOD) in the late 1980s, and TOD became a fixture of modern planning in the US (Calthorpe, 1993). In 1961, Jane Jacobs published *The Death and Life of Great American Cities* and coined the terms "social capital", "mixed primary uses", and "eyes on the street", phrases that are commonly used in urban design, sociology, and other fields today.

Half a century later, most cities in the world are still grappling with issues concerning sustainable transportation and safety on urban streets. The more complex the issue, the greater is the tendency to focus on the simple and technical aspects of the problems. Sustainable transport solutions are frequently reduced to those concerning cleaner vehicle emissions, provision of public transport, and "encouraging" walking and bicycling. The reasons why people and governments do not or cannot follow many of the prescribed goals get less attention. Managing transport and providing access to all citizens in a city is one of the most profound and important challenges facing global society in a future without the plentiful and cheap oil resources based on which modern urban economies have been built.

This volume includes contributions of an international interdisciplinary group of experts who met in Agra (India) to discuss the challenges facing us in the design of urban structures, policies and technologies and to suggest possible ways forward. The workshop was sponsored by the Volvo Research and Educational Foundations, Infrastructure Development Finance Company (India) and the Ministry of Urban Development, Government of India. The contributions focus on forms of city structure, urban governance, alternate transport policies, safety and the environment, paratransit operations and emerging technologies. A special feature of these contributions and discussions was the focus on recent knowledge generation of ideas and data from cities in low- and middle-income countries.

The organisation of urban governance mechanisms that deal with transport has attracted a great deal of attention in recent years, and in his chapter Gerald Frug captures the complexity of the problem succinctly:

> Transportation decisions will be illegitimate unless they respond to local residents' needs and desires – respond, in other words, to the democratic will. But this does not mean that there isn't a role for experts too. Of course experts have a role. But expert decisions have to be connected to democratic decisions. In short, we need to be broad and narrow at the same time. This conventional wisdom is easy to state. But, as far as I know, it is not implemented anywhere in the world.

In addition, Dunu Roy suggests that, in a world where inequality is increasing in many locations, these complexities make the implementation of people-friendly transport policies even more difficult:

> There appears to be a deepening gap between the institutions of government and the people ... It is clear that one set of stakeholders, among the many, is calling the shots, and is railroading the rules to strengthen its convenience, its profits, and its control.

Taking governance issues as a given, Hermann Knoflacher gives suggestions for future research but maintains that the sustainable development of cities would be possible only if planners make parking regulations that separate parking sites from human activities. According to him if they are not able to implement and monitor these principles they would only be treating the symptoms and not taking care of the disease. Though many problems confront us, A. G. Krishna Menon informs us that:

> In terms of settlement density, social heterogeneity, and economic mix, traditional cities are examples of an urban character not seen in other parts of the city developed by modern town planners. While they have real problems, such as infrastructural and other forms of deprivation, as urban typology they are both satisfying and appropriate models for imagining an Indian city.

Sirish B. Patel and Philipp Rode deal with city design densities and transport planning. Rode's work from London demonstrates that only density levels of more than 100 persons per hectare ensure that the percentages of public transport trips to work remain above 40% and he suggests that this would offer clues "that might become increasingly relevant for a developing world context". However, Patel's data from Mumbai suggest that densities there are generally much greater than 100 persons per hectare and that higher densities should not be confused with high-rise buildings and higher floor space index (FSI). He warns us that, with higher density of occupants in dwelling units in low- and middle-income countries, it is much more important to focus on overall density per hectare than FSI regulations. Otherwise, there simply will not be space on the streets for people to move around.

In addition to concerns with city governance and urban form, the participants also focused on issues of alternative transport policies, influence of city infrastructure on access, role of paratransit, taxis and non-motorised transport, safety and technology. Geetam Tiwari warns us that:

> There is a conflict of meanings of sustainability between the rich and the poor. For the rich sustainability can be achieved thorough technological fixes – cleaner engines, fuels and metro systems. For the poor, sustainability is a lifestyle and employment issue more addressed by city form, land use and accessibility issues. This conflict manifests itself politically, with no easy or fast solutions.

It is in this context that Elliott Sclar provides us details of attempts to create a planning context for implementing a public transport system in Nairobi (Kenya); and Arif Hasan informs us about the role the Urban Resource Centre played in creating a space for consensus-building in Karachi (Pakistan) between various actors in the urban drama so that communities and weaker interest groups could regain ownership of the city.

As far as actual mobility is concerned "informal" transport services (minibuses, vans, taxis, three-wheelers, motorcycle taxis) transport more people than formal public transport systems in a large number of cities around the world. Robert Cervero, Geetam Tiwari, and Roger Behrens provide data and case studies from around the world that show that these services may be termed "informal" but they often function within reasonably structured systems and also comply with local regulations imposed by local authorities. The availability of these systems at the relatively low fares they charge makes it possible for millions of people to access their jobs and their daily needs. Roger Behrens concludes:

> Indeed, an overarching theme across all the modes discussed in this paper is the absence of adequate attention in policy debates to "softer" institutional and implementation issues. Progress in resolving these issues will prove central to the creation of more sustainable non-motorised, for-hire and paratransit modes.

While urban transport planning issues dominate the discourse at policy and technical levels, pollution and road safety concerns agitate the larger community also. Unsafe roads and neighbourhoods contribute to a greater use of personal motorised transport and increase pollution. In his contribution, Matthijs J. Koornstra uses evidence from Europe and the United States to show the relationship between safety and various factors including speed, policing intensity and vehicle design. However, he is convinced that "The major obstacles for the zero vision achievement of a sustainable safe road traffic are the lack of sufficient political priority for road safety, even in highly motorised countries, and the general ignorance of the existence of effective road safety measures by authorities". In his chapter, Dinesh Mohan endorses the issues raised by Matthijs J. Koornstra and adds that, to promote walking, bicycling and public transport use, we will have to make traffic safety a priority along with city designs that include denser layouts combined with smaller size of city blocks. Carlos Dora brings together the issues of traffic safety, health and the built environment, and climate change, and insists that evaluation of the performance of transport policies should involve documentation of all the health impacts those policies have on city populations.

The contributors to this volume bring to us the depth and diversity of the challenge we face in the way cities are conceived, adapted, designed, developed, and managed in a world confronted with increasing road accident fatalities, and disease and death due to climate change and atmospheric pollution. These contributions should help policy makers and researchers to better equip themselves for the road ahead.

Dinesh Mohan
Geetam Tiwari
July 2018

REFERENCES

Calthorpe, P. (1993) *The Next American Metropolis: Ecology, Community, and the American Dream.* Princeton, NJ: Princeton Architectural Press.
Carlton, I. (2007) Histories of Transit-Oriented Development: Perspectives on the Development of the TOD Concept: Real Estate and Transit, Urban and Social Movements, *Concept Protagonist.* Berkeley, CA: Institute of Urban and Regional Development University of California, Berkeley.

Editors

Dinesh Mohan is Honorary Professor at the Indian Institute of Technology Delhi, India, and Director of the Independent Council for Road Safety International. For many years, he was a member of the World Health Organisation (WHO) Advisory Panel on Accident Prevention. He serves on the editorial boards of four international journals dealing with safety. He has been a consultant on safety-related matters to government departments in India, Nepal, Indonesia, Thailand, Bangladesh, Iraq, and Libya; and automotive industries including TELCO, Ashok Leyland, Volvo Trucks, Eicher Motors, Escorts, Maruti Udyog, SIAM, and Bajaj Auto; and also to international organisations like the World Bank and the WHO. His research includes the following areas transportation research (safety and pollution), human tolerance biomechanics, motor vehicle safety, road traffic injuries, effectiveness of automobile safety equipment, evaluation of injuries to cyclists and motorcyclists, motorcycle helmet design, and evaluation of governments' and motor-vehicle manufacturers' standards concerning motor vehicle safety. He is the recipient of (1) a Distinguished Career Award from the University of Delaware, Newark, DE; (2) Distinguished Alumnus Award 2002 from the Indian Institute of Technology, Bombay, India; (3) International Research Council on Biokinetics of Impacts' 2001 Bertil Aldman Award for Outstanding Research on the Biomechanics of Impacts; (4) the American Public Health Association International Distinguished Career Award in recognition of dedication and leadership in the area of injury research and teaching, with contributions and achievements that have significant and long-term impact on the problem of injury prevention and control; (5) the International Velo-City Falco Lecture Prize; (6) the Association for the Advancement of Automotive Medicine's 1991 Award of Merit for outstanding research in traffic safety; and (7) the 1991 International Association for Accident & Traffic Medicine's International Award and Medal for Outstanding Achievement in the Field of Traffic Medicine.

Geetam Tiwari obtained her BArch from the University of Roorkee, India, and Master of Urban Planning and Policy and PhD in Transport Planning and Policy from the University of Illinois, Chicago, IL. Currently she is MoUD Chair Professor for Transport Planning at the Indian Institute of Technology Delhi, India. She was Adlerbretska Guest Professor for sustainable urban transport at the Chalmers University of Technology, Sweden, from 2007 to 2010. She has extensive research experience in dealing with transportation issues of special relevance to low-income countries. These include development of bus systems and road designs that would make transportation efficient and safer. She has been working in the area of traffic and transport planning, focusing on pedestrians, bicycles, and bus systems. She has published over 70 research papers on transportation planning and safety in national and international journals and peer-reviewed seminar proceedings, as well as edited four books on transportation planning and road safety. She received the International Velocity Falco Lecture Prize, Barcelona, Spain, the Stockholm Partnerships award for local impact, innovative thinking, and potential for replication or transferability, and the Centre for Excellence grant from Volvo Research and Educational Foundations (VREF), IRTE, and Prince Michael's award for promoting road safety research and LMA (Lucknow Management Association) award for woman achiever, 2010. She is advisor to the Urban Age series of conferences coordinated by the London School of Economics, UK, since 2005. Since 2009, she has been editor-in-chief of the *International Journal of Injury Control and Safety Promotion*.

Contributors[1]

Roger Behrens is Convenor of the postgraduate Transport Studies Programme of the Faculty of Engineering and the Built Environment, University of Cape Town (UCT), South Africa, and a founding member of the multidisciplinary Urban Transport Research Group. He is Senior Lecturer in the Department of Civil Engineering. He graduated with an MA in City and Regional Planning (with distinction) from UCT in 1991 and with a PhD in 2002. In 1996, he became a corporate member of the South African Planning Institution, and a registered professional Town and Regional Planner. He has 15 years of experience in research and in the management of commissioned research projects. His core research and consultancy experience is in travel surveys and travel behaviour, site layout planning and neighbourhood movement networks, growth management, and low-income housing and infrastructure provision. He serves on the Transport Appeal Tribunal, which considers appeals related to the issuing of operating licenses to paratransit minibus-taxi operators. He has been invited to chair conference sessions, and act as a conference rapporteur. He referees international journal articles. His papers have been published in international peer-reviewed journals, he has authored chapters in books, and he has presented peer-reviewed papers at numerous international and national conferences. His current research is on non-motorised transportation, changing travel behaviour, scholar travel, and policy analysis of urban passenger transport and land use-transport interaction. He teaches courses on transport system supply and demand management, local area transport planning and management, and non-motorised transportation.

Ricky Burdett is a London-based urban specialist with a wide portfolio of academic and consultancy activities at an international scale. Trained as an architect, Burdett is a recognised world authority in urban development and design, contemporary architecture, and the social and spatial dynamics of contemporary cities. He leads LSE Cities, a global centre of research and teaching at the London School of Economics and Political Science which

---

[1] Details as at date of contribution.

received The Queen's Anniversary Prize for Higher and Further Education in 2018 for its work on 'Shaping Urban Leaders and Cities of the Future'. He founded and directs the authoritative Urban Age programme, an inter-disciplinary investigation of global cities that brings together national and city leaders, academics, designers and civic actors.

Alongside his academic activities, Professor Burdett acts as a consultant to national and city governments, private companies and philanthropic agencies. He advises investors and developers as well as public bodies on improving the quality of design and the environment, developing archi-tectural and urban design briefs, running and assessing design procure-ment processes including architectural and infrastructure competitions. He has played a key role in high-profile urban projects like the London 2012 Olympics and legacy, the Tate Modern and new buildings for the BBC, and advises private and public organisations in Europe, USA, the Middle East and Asia on a range of urban developments at varying scale.

Professor Burdett is a regular commentator in the media on contempo-rary architecture and cities, and a keynote speaker at major conferences and institutions including UN Habitat and the World Economic Forum in Davos. He was Director of the Venice International Architecture Biennale and Curator of the Global Cities Exhibition in the Turbine Hall of Tate Modern in London. He was a member of the UK Government Airport Commission and Council member of the Royal College of Art, is a Trustee of the Norman Foster Foundation and acts a Cultural Ambassador as a member of the Mayor of London's Cultural Leadership Board. He sits on the jury for Royal Academy Architecture Prize and Dorfman Awards, and is chair of the Mies Crown Hall Americas Prize 2018.

Professor Burdett was appointed a CBE (Commander of the Order of the British Empire) for services to urban planning and design in the 2017 New Year's Honours List and was awarded an Honorary Fellowship by the Royal College of Art in 2019.

Fabio Casiroli was born in Milan in 1950, and is a freelance specialist who has been working in the field of territorial and transport planning since 1975.

He's been a contract Professor in Transport Planning since 1998 in the Faculty of Civil Architecture at the Polytechnic of Milan, Italy. He has held contract professor positions in the Faculty of Engineering at the University of Cagliari, Italy, and the Faculty of Architecture in Palermo, Sicily, and has taught at IUAV (Istituto Universitario di Architettura di Venezia) in Venice, Italy. He has participated in research projects for universities, the CNR (National Center for Research), Italy, the European Union, and the Banco Interamericano de Desarrollo, US.

In 1989 he founded Systematica, a public limited company, of which he is chairman, that offers consultancy services, territorial and transport planning, traffic engineering and design services. He has worked in Europe,

Latin America, Africa, and Asia on studies, plans, and projects on both an urban and a regional scale, where applied techniques for evaluating the interventions were supplied by advanced simulation codes and sophisticated methods of demand analysis. He has led or been the specialist consultant on the design of dozens of urban traffic plans as well as specialist studies for regional and provincial transport plans, railway and subway systems, goods hubs, road and motorway infrastructures, and traditional and innovative transport services. For several years he has provided specialist consultancy for some of the most notable contemporary architects on large projects in Italy and abroad. He has curated the "traffic and mobility" section of the Venice Architecture Biennale 2006, where he presented studies on 12 world megacities.

Harry T. Dimitriou has been Bartlett Professor of Planning Studies at the Bartlett School of Planning, University College London (UCL), UK, since 1998, and was its head of school between 1998 and 2004. He is currently director of the OMEGA Centre – a global centre of excellence for the study of mega urban transport projects at UCL funded by the Volvo Research and Education Foundations (VREF). He holds a Diploma in Town and Regional Planning from Leeds School of Town Planning, an MSc in Urban Science from University of Birmingham, UK, and a PhD in Transport and Urban Development from the University of Wales, UK.

He has held numerous overseas advisory positions in city and regional planning, transportation policy-making and planning, and in institution-building in Greece, China, India, Indonesia, Jordan, Nigeria, and Saudi Arabia – for the European Commission, the World Bank, the United Nations Development Programme, the United Nations Centre for Human Settlements, the Economic and Social Commission for Asia and the Pacific, the Harvard International Development Institute, the Hong Kong government, the Lagos state government, and the government of Indonesia. He has been a member of the Transport Working Group for the Intergovernmental Panel on Climate Change; UK Government Office of Science and Technology Foresight Built Environment and Transport Panel; Committees of the Transportation Research Board (TRB) in Washington, DC, on socioeconomic aspects of transportation and transport in developing countries, and is currently member of the UK Government Task Force (nCRISP) examining research in major projects.

Carlos Dora is an environmental health policy expert with WHO, where he leads the development of new approaches to impact assessment that include both environment and health. He presently is involved in a series of pilot projects testing new impact assessment approaches in the developing countries of Africa and Asia. As a medical doctor he first worked to develop quality primary health care in Latin America. He later completed a PhD at the London School of Hygiene and Tropical Medicine, UK, investigating diet and non-communicable diseases, with a developing country emphasis. He then moved to the WHO's European Centre for Environment and Health, Germany, where he dealt with a range of environmental health issues, including the health impacts of Chernobyl and depleted uranium and capacity-building in environmental epidemiology. He then went on to create a new WHO programme on the health implications of transport policies and contributed to developing an intergovernmental plan of action for healthy transport. He has since led the development of the health aspects of a new Protocol on Strategic Environment Assessments, to the Espoo Convention on Environmental Impact Assessment. A former senior policy analyst for the WHO Director General's office, his research work included models for HIA of transport scenarios, and the interplay of mass media, government and health care discourses in risk communication strategies.

Gerald Frug is Louis D. Brandeis Professor of Law at Harvard Law School, US. Educated at the University of California at Berkeley, US, and Harvard Law School, he worked as a Special Assistant to the Chairman of the Equal Employment Opportunity Commission, in Washington, DC, US, and as Health Services Administrator of the City of New York, US, before he began teaching in 1974 at the University of Pennsylvania Law School, US. He joined the Harvard faculty in 1981. His specialty is local government law, a subject he has taught for more than 25 years. He has published dozens of articles on the topic and is the author, among other works, of a casebook on *Local Government Law* (4th edition, West Academic, 2006, with David Barron and Richard T. Ford), *Dispelling the Myth of Home Rule (Rappaport Institute*, 2004, with David Barron and Rick Su), and *City Making: Building Communities without Building Walls* (Princeton University Press, 1999).

Arif Hasan is an architect/planner in private practice in Karachi, Pakistan. He studied architecture at the Oxford Polytechnic, UK, and, on his return to Karachi in 1968, established an independent practice, which slowly evolved into dealing with urban planning and development issues. He has been a consultant and advisor to many local and foreign CBOs, national and international NGOs, and bilateral and multilateral donor agencies. Since 1982, he has been involved with the Orangi Pilot Project (OPP) and is the founder Chairman of the Urban Resource Centre (URC), Karachi, since its inception in 1989. The OPP is an informal settlement upgrading project whose development is managed and funded by local communities. The URC is a research and advocacy organisation supporting communities against eviction and against gentrification and/or degradation of Karachi's inner city.

He has taught at Pakistani and European universities, served on juries of international architectural and development competitions, and has authored numerous books. He was a celebrity speaker at the Union of International Architects Congress in Brighton in 1987 and has been a member of the Steering Committee of the Aga Khan Award for two cycles. He is currently on the board of several international journals. He has received a number of awards for his work including the UN Year for the Shelterless Memorial Award of the Japanese Government (1990), the Prince Claus Award of the Netherlands Government (2000), and the Hilal-i-Imtiaz of the Government of Pakistan (2001). Recently, he was given a Life Time Achievement Award by the Institute of Architects, Pakistan (2003). The Orangi Project Research and Training Institute, of which he is Chairman, received the British Housing Foundation's World Habitat Award in 2002.

Hermann Knoflacher obtained his PhD in Technical Science from the Technical University of Vienna, Austria. Currently he is University Professor and head of the Institute for Transport Planning and Traffic Engineering, University of Technology Vienna (TU Wien), Austria. He has been Head of the Institute for Transport, Austrian Road Safety Board. His main research focus includes design of transport elements; transport system user behaviour; traffic infrastructure and mobility; sustainable development of cities and mobility; traffic safety; energy consumption; environment; and basic interdisciplinary research. He has published five books and more than 400 scientific publications.

Matthijs J. Koornstra studied psychology and computational mathematics and graduated in psychology at Leiden University, the Netherlands, where he worked from 1961 to 1986, the first five years as research assistant, the second five years as researcher at the Computer Centre and the Psychological Institute. He also worked as a senior researcher at the Department of Data Theory of the Faculty of Social and Behavioural Sciences from 1971 to 1978, and from 1978 to 1986 as (crown-appointed) vice-president of Leiden University. In 1986 he was appointed as director of SWOV (Research Institute for Road Safety, the Netherlands) and after 1999 as director of SARA (National Centre for Advanced Supercomputing and Networking, Amsterdam, the Netherlands). Meanwhile he lectured on road safety for ten years at the Delft University of Technology, the Netherlands, and was or is president or member of governing boards for scientific foundations, academic visiting committees, and (inter)national scientific councils. Since 2002 he worked independently as research advisor for various international organisations (EU, ETSC, OECD, ECMT, WB, WHO, and WBCSD). He is author of the book *Changing Choices: Psychological Relativity Theory* (Leiden University Press, 2007) on judgment, preference, and risk behaviour; co-editor or co-author of several other books on road safety; and author or co-author of over 120 scientific articles or reports on data analysis methods, road safety, and many diverse domains in applied and theoretical psychology, meanwhile obtaining his PhD at Leiden University.

A. G. Krishna Menon is an architect, urban planner, and conservation consultant practising in Delhi for over 30 years, while at the same time teaching in Delhi. In 1990 he co-founded the TVB School of Habitat Studies in New Delhi, India, and is currently its director. He is actively engaged in research, and has contributed extensively to professional journals and several academic books. He has been actively involved in urban conservation and has been associated with the formulation of The Delhi Master Plan – 2021, The National Capital Region Master Plan – 2021, and is a member of several statutory committees set up by the government.

Haixiao Pan has been Director of Land Use/Transport Studies in the Department of Urban Planning, Tongji University, China, since 1996. He has been a Board Member of Shanghai Urban Economics Institute and holds a PhD from Shanghai Jiaotong University, China (1989), and Diploma in Planning from the University of Sheffield, UK. His recent work includes major master planning and transportation planning exercises including The Transport Management Framework for Shanghai 2010 Expo; The Transport Strategy Study for Shengyang and Zibo City; The Impact of the China Coastal Freeway on the Development of Shanghai; The Integration of Land Use and Transport Planning System Study; The Master Plan of Laiwu City in Shandong Province; The Regulatory Plan of the New District; and The Urban Transportation Plans of Yueqing City (Zhejiang Province), Linan City (Zhejiang Province), Shaoxing City, the Baoshan New District of Shanghai, and the Qingpu District of Shanghai.

Shirish B. Patel founded a firm of Consulting Civil Engineers in 1960 that has undertaken design and supervision of construction of a wide variety of civil works all over India, including road and rail bridges, railway stations, and elevated rail-track. He has personally had a long and sustained interest in urban affairs. He was one of the three original authors who suggested the New Bombay project in 1965. In 1970, when the government accepted the project and set up City and Industrial Development Corporation (CIDCO), the new town authority, he was appointed in charge of all planning, design and execution, a position he left five years later because he was unhappy with the lack of sustained political interest in the project. He was then a Member of the Executive Committee of the Bombay Metropolitan Regional Planning Board, the overall planning authority for the region, for 13 years. Later, his firm, in association with Rahul Mehrotra Associates, carried out the work of preparing the Development Plan for the southern half of the Vasai-Virar Region in northern Mumbai, which included a scheme for making the infrastructure development self-financing. He has been a member of the Mumbai Heritage Conservation Committee, and a governor on the board of the Mumbai Metropolitan Region Heritage Conservation Society. He has retired from the consultancy firm he founded, of which he is now

Chairman Emeritus, and devotes his time to work on urban development and urban affairs.

Philipp Rode is Executive Director of London School of Economics (LSE) Cities and Associate Professorial Research Fellow at the London School of Economics and Political Science. He is Co-Director of the LSE Executive MSc in Cities and co-convenes the LSE sociology course on 'City Making: The Politics of Urban Form'. As researcher, consultant and advisor he has been directing interdisciplinary projects comprising urban governance, transport, city planning and urban design at the LSE since 2003.

The focus of Rode's current work is on institutional structures and governance capacities of cities, and on sustainable urban development, transport and mobility. He recently published 'Governing Compact Cities: How to connect planning, design and transport' (Elgar 2018), co-edited 'Shaping Cities in an Urban Age' (Phaidon 2018) and co-authored 'Resource Urbanisms: Asia's divergent city models of Kuwait, Abu Dhabi, Singapore and Hong Kong' (2017), and 'Towards New Urban Mobility: The case of London and Berlin' (2015).

Rode is Executive Director of the Urban Age Programme, manages its global research efforts and leads on its latest task force advisory initiative for the city of Addis Ababa. Since 2005, he has organised Urban Age conferences in over a dozen world cities, bringing together political leaders, urban practitioners, and academic experts. He co-directed the cities workstream of the Global Commission on the Economy and Climate in the run-up to the 2015 UN Climate Change Conference in Paris and co-led the UN Habitat III Policy Unit on Urban Governance. He has previously led the work on green cities and buildings for the UN Environment Programme's Green Economy Report.

Rode is a Member of the Board of Directors of the Institute for Transportation and Development Policy (ITDP) and Steering Committee Member of the Coalition for Urban Transitions led by the C40 Cities Climate Leadership Group and the World Resources Institute. He was awarded the Schinkel Urban Design Prize in 2000.

A. K. (known more usually as Dunu) Roy trained as a chemical engineer, forced into social science by compulsion, political ecologist by choice. He has spent almost four decades interacting with community groups in rural and urban India, listening mostly, occasionally providing information and advice on how to deal with problems and several years experimenting with failure in Shahdol district of Madhya Pradesh, central India, trying to understand society, technology, and the relationship between the two. He has also studied wild life, pollution, development, watersheds, and other assorted forms of human vagaries, and has provided consultancies on thermal power, water management, coal mining, highways, energy management, land reforms, disaster management, and so on, with clients such as the World Bank, GTZ, Oxfam, and Dorabji Tata Trust – institutions with much input but little insight. He occassionally serves as research advisor for the People's Science Institute at Dehradun, India. He currently happily occupies a directorial chair in the Hazards Centre at Delhi, India.

Elliott Sclar is Director of CSUD and Professor of Urban Planning and Public Affairs at Columbia University, US. He holds senior appointments in the Graduate School of Architecture, Planning, and Preservation and the School of International and Public Affairs. He is Director of graduate programs in Urban Planning and is an active participant in the work of the Earth Institute at Columbia University, EI. He was the co-coordinator of the Taskforce on Improving the Lives of Slum Dwellers. It is one of the ten taskforces set up by the UN Millennium Project to help guide the implementation of the United Nation's Millennium Development Goals. The Taskforce's book length report (2005), "A Home in the City" (PDF download), is available on the UN Millennium Project website and from Earthscan.

As a professional economist, he has written extensively about the strengths and limitations of markets as mechanisms for effective public policy implementation. The main focus of this Center will be to work at the nexus that connects the regulatory mechanisms of planning with market-based incentives to create environmentally and economically sustainable urban development. His book *You Don't Always Get What You Pay For: The Economics of Privatization* (Cornell University Press, 2000), a critique of overreliance on market mechanisms, has won two major academic prizes: the Louis Brownlow Award for the Best Book of 2000 from the National Academy of Public Administration and the 2001 Charles Levine Prize from the International Political Science Association for a major contribution to the public policy literature. It is a definitive work in the field.

K. C. Sivaramakrishnan has been Honorary Visiting Professor at the Centre for Policy Research, New Delhi, India, since 1996 and Senior Fellow, Institute of Social Sciences since 2003. Presently he is also chairman of the CPR board.

After joining the Indian Administrative Services (IAS) in 1958 and holding various assignments in West Bengal, he moved to Delhi in 1985 to become the first project director of the Central Ganga Authority and additional secretary in the Ministry of Environment. In 1988, he became secretary in the Ministry of Urban Development. In that capacity, Sivaramakrishnan was personally involved in the legislation to amend the Constitution to provide a framework for decentralisation and empowerment of rural and urban local bodies. Eventually the 73rd and 74th Amendments became part of the Constitution.

Sivramakrishnan retired from the IAS in 1992. Thereafter he joined the World Bank as senior advisor, Urban Management. Since his return to the country in 1996 Sivaramakrishnan has been with the Centre for Policy Research as professor and is also associated with the Institute of Social Sciences, Delhi, and Lok Satta, a movement for electoral reforms.

In the academic sphere Sivaramakrishnan was a Pravin fellow at Woodrow Wilson School, Princeton University, US, in 1965, visiting professor and Homi Bhabha fellow at the Indian Institute of Management, Calcutta, India, in 1977, and senior lecturer at the Economic Development Institute of the World Bank from 1978 to 1982.

Trained in economics, political science and law, he has authored several books and papers on urban management, decentralisation, and urban environment.

Julie Touber received her MSc in Urban Planning from the Graduate School of Architecture, Planning and Preservation at Columbia University, US. She also holds an equivalent planning degree from La Sorbonne in Paris, France. She has extensive experience in the developing world. Following an internship at the World Heritage Center at UNESCO, Paris, France, she wrote her first thesis on the impacts of preservation policies on World Heritage Sites based on the case of Luang Prabang in Laos. She was part of the Columbia University, US, urban planning studio team that was sent to Accra, Ghana, to study disaster mitigation. Following this experience, she wrote a paper on the "Sanitation Crisis in Accra, Ghana". Her Master's thesis related to the scale of planning practices and development policies in the developing world, using the case study of Sana'a in Yemen.

Abbreviations

ABM	American business model
ACCP	Action Committee for Civic Problems
ACHR	Asian Coalition for Housing Rights
ADB	Asian Development Bank
AMC	Ahmedabad Municipal Corporation
AMIS	air management information system
AMTS	Ahmedabad Municipal Transport Service
APA	American Planning Association
APSRTC	Andhra Pradesh State Road Transport Corporation
BAC	blood alcohol content
BLC	Bhure Lal Committee
BLIP	bus lanes with intermittent priority
BMJ	British Medical Journal
BOM	Mumbai
BOOT	Build Own Operate Transfer
BOT	Build-Operate-Transfer
BRT	bus rapid transit
BRTS	bus rapid transport system
BUA	built-up area
C	car
C&T	car and transit
CB	Cantonment Board
CB	citizen's band
CBD	central business district
CBO	community-based organization
CBPA	commercial buildable plot area
CBUA	commercial built-up area
CD8	community district 8
CEC	Commission of the European Communities
CFMT	Citizens' Forum on Mass Transit
CIDA	Canadian International Development Authority
CL	Central London
CLUP	comprehensive land-use plan

CMS	changeable message signs
CNG	compressed natural gas
COG	Council of Local Governments
COPD	chronic obstructive pulmonary disease
CPCB	Central Pollution Control Board
CRRI	Central Road Research Institute
CSE	Centre for Science and Environment
CSUD	Center for Sustainable Urban Development
DDA	Delhi Development Authority
DEL	Delhi
DG-TREN	Directorate General for Energy and Transport
DHA	Defence Housing Authority
DIMTS	Delhi Integrated Multimodal Transport System
DMRC	Delhi Metro Rail Corporation
DOT	Department of Transport
DRL	daytime running lights
DRT	demand responsive transport
DTA	Delhi Transport Authority
DTC	Delhi Transport Corporation
DTS	Delhi Transport Services
DTU	Delhi Transport Undertaking
DU	dwelling unit
DWI	driving while intoxicated
ECMT	European Conference of Ministers of Transport
EIA	environment impact assessment
ETH	Eidgenössische Techische Hochschule
EU	European Union
FAR	floor area ratio
FSI	floor space index
FTA	Federal Transit Administration
GB	Great Britain
GDP	gross national product
GHG	greenhouse gases
GIS	geographic information system
GL	Greater London
GLUC	generalised land-use classification
GNCTD	Government of National Capital Territory of Delhi
GOI	Government of India
GPS	global positioning system
GTZ	Gesellschaft für Technische Zusammenarbeit
HCM	*Highway Capacity Manual*
HCV	heavy commercial vehicle
HDFC	Housing Development Finance Corporation
HEARTS	health effects and risks of transport
HIC	high-income country

HMC	highly motorised country
HOU	Houston
HOV	high-occupancy lane
IATSS	International Association of Traffic and Safety Sciences
IBRD	International Bank for Reconstruction and Development
ICICI	Industrial Credit and Investment Corporation of India
ICT	information and communications technology
ICTSL	Indore City Transport Services
IIED	International Institute for Environment and Development
IFC	International Finance Corporation
IFI	international financial institution
IIT	Indian Institute of Technology
IL	Inner London
ILFS	infrastructure leasing and financial services
IMD	index of multiple deprivation
INTACH	Indian National Trust for Art and Cultural Heritage
IPCC	Intergovernmental Panel on Climate Change
IPT	intermediate public transport
ISBT	interstate bus terminal
ISTEA	Intermodal Surface Transportation Act
IT	information technology
ITDP	Institute for Transportation Development and Policy
ITPI	Institute of Town Planners, India
ITS	intelligent transport systems
ITS Surabaya	Sepuluh Nopember Institute of Technology
IUCN	International Union for Conservation of Nature
JAMA	Journal of American Medical Association
JNNURM	Jawaharlal Nehru National Urban Renewable Mission
JOS	Jamaica Omnibus Service
JSTOR	Journal Storage
JUTC	Jamaica Urban Transport Company
KDA	Karachi Development Authority
KIPPRA	Kenya Institute for Public Policy Research and Analysis
KMC	Karachi Metropolitan Corporation
KMC	Kingston Metropolitan Area
LCV	light commercial vehicle
LG	Lieutenant Governor
LMIC	low- and middle-income countries
LOV	low occupancy vehicle
LPG	liquefied petroleum gas
LRT	light rail transit
LSE	London School of Economics and Political Science
LNWA	Lyari Nadi Welfare Association
MC	municipal corporation
MCD	Municipal Corporation of Delhi

MCV	medium commercial vehicle
MDG	Millennium Development Goals
MMRDA	Mumbai Metropolitan Region Development Authority
MNA	Members of the National Assembly
MoUD	Ministry of Urban Development
m/p	meters per person
MPC	Metropolitan Planning Committee
MPD	Master Plan for Delhi
MPD-2001	Master Plan for Delhi – 2001
MPD-2021	Master Plan for Delhi – 2021
MRTS	mass rapid transport system
MTW	motorized two-wheeler
MV	motor vehicle
NCAP	New Car Assessment Programme
NCD	non-communicable disease
NED	NED University of Engineering and Technology
NGO	non-governmental organisation
NDMC	New Delhi Municipal Council
NMIV	Nichtmotorisierter Individualverkehr
NMT	non-motorised transport
NMV	non-motorised vehicle
NOIDA	New Okhla Industrial Development Authority
Nox	nitrogen oxides
NUTP	National Urban Transport Policy
NYC	New York City
NY-CD	New York community district
ODPM	Office of Deputy Prime Minister
OECD	Organisation for Economic Co-operation
OL	Outer London
OPP	Orangi Pilot Project
PAK	Pakistan
pcu/hr	passenger car units per hour
PFI	Private Finance Initiative
PGA	public ground area
PHT	people hours trip
PIL	public interest litigation
PKT	people kilometers trip
PM	particulate matter
pphpd	passengers per hour per direction
PPP	purchasing power parity
PTAL	Public Transport Accessibility Level
PWD	Public Works Department
RBPA	residential buildable plot area
RBUA	residential built-up area

RCMRD	Regional Center for Mapping of Resources for Development
RITES	Rail India Technical and Economic Service
RMB	Renminbi
RTA	road traffic accident
RTI	road traffic incidents
RTI	road traffic injury
SAFETEA-LU	Safe, Accountable, Flexible Efficient Transportation Equity Act – Legacy for Users
SARS	severe acute respiratory syndrome
SC	Supreme Court
SHI	Secure Housing initiative
SO$_2$	sulphur dioxide
SPM	suspended particulate matter
SUN	Sweden, the United Kingdom, the Netherlands
SW	Sweden
SWOV	Institute for Road Safety Research in the Netherlands
T	transit
TCPA	Town and Country Planning Association
TM	thematic mapper
TN	Tamil Nadu
TNUDF	Tamil Nadu Urban Development Fund
TOD	transport-oriented development
tpd	trips per day
TRIPP	Transportation Research and Injury Prevention Programme
TRL	Transport Research Laboratory
TSR	three-wheeled scooter rickshaw
TV	television
TVB	TVB School of Habitat Studies
TWU	Transportation, Water and Urban Development Department
UA	urban agglomeration
UCB	University of California, Berkeley
UDPFI	Urban Development Plans Formulation and Implementation
UK	United Kingdom
US	United States
USA	United States of America
USD	United States Dollar
UMTA	Unified/Urban Metropolitan Transport Authority
UNDP	United Nations Development Programme
URC	Urban Resource Centre
UT	Union Territory
vht	vehicle hours trip

vkt	vehicle kilometers trip
VTI	National Road and Transport Research Institute, Sweden
WB SSATP	World Bank, Sub-Saharan African Transport Plan
WBCSD	World Business Council for Sustainable Development
WHO	World Health Organization
WMO	World Meteorological Organization
WTO	World Trade Organization

Chapter I

Urban transportation planning

Geetam Tiwari

CONTENTS

INTRODUCTION

Urban transport systems and city patterns have a natural interdependency. Land-use patterns, population density and socioeconomic characteristics influence the choice of transport system; at the same time the presence of certain transport systems changes land accessibility and therefore land value, triggering a change in land-use pattern and city form. Therefore city and transport plans and policies prepared by the policy makers, experts and decision makers are expected to play an important role in influencing the future health of our cities. However, a large proportion of the urban populations in Asian and other low-income cities remain outside the formal planning process. Survival compulsions force them to evolve as self-organised systems. These systems rest on the innovative skills of people struggling to survive in a hostile environment and to meet their mobility and accessibility needs. Housing, employment and transport strategies adopted by this section of society are often termed as "informal housing, informal employment and informal transport". Squatter settlements all over the world are called informal settlements because they are not part of the official plan. The conventional definition of informal – unofficial, illegal or unplanned – denies people jobs in their home areas and denies them homes in the areas where they have gone to get jobs. Transport solutions evolved by this section of society do not become part of official policy. Their existence is mostly viewed as creating problems for "normal traffic". Formal plans have no place for informal transport. Therefore, most cities face a complex situation where investments are for formal plans, whereas the needs of a significant section of society are met by informal transport. Is this desirable or sustainable?

In this chapter, we discuss the processes of transport policy and planning in Indian cities and compare them with examples from some other cities around the world. Policies and plans adopted by different cities in different regions of the world have a striking similarity. Official plans for Mexico city, Shanghai or Delhi may be intended for different countries, and different regions, but the disconnect between the policy documents that emphasize sustainable transport solutions, recognizing the environment and health problems of the citizens, and the projects approved for construction that primarily address the needs of car users and promote large construction projects, have strong similarities across the world.

URBAN TRANSPORT IN INDIAN CITIES

Indian cities of all sizes face a crisis of urban transport. Despite investment in road infrastructure and plans for land use and transport development, all cities face the problems of congestion, traffic accidents and air and noise pollution. All these problems are on the increase. Large cities are facing a rapid growth in personal vehicles (two-wheelers and cars), and in medium and small cities different forms of intermediate public transport provided by the informal sector are struggling to meet the mobility demands of city residents. Several attempts have been made by planning authorities and experts to address these problems. The Rail India Technical and Economic Service (RITES) has summarised studies related to urban transport policy since 1939 (RITES, 1998). A national urban transport policy was adopted by the Ministry of Urban Development in March 2006 (MoUD, 2006). Land-use master plans prepared for most metropolitan cities have a brief chapter on urban transport. Urban transport has been given a major emphasis in the Jawaharlal Nehru National Urban Renewable Mission (JNNURM) by offering financial assistance to the state and city governments in order to prepare comprehensive traffic and transport studies and grant assistance for implementing pedestrian, non-motorised transport and public transport friendly infrastructure. The government has repeatedly made statements regarding its commitment to providing safe and sustainable transport for the masses. However, investments in road widening schemes and grade-separated junctions, which primarily benefit personal vehicle users only (car and two-wheeler users), have dominated its expenditure. The total funds allocated for the transport sector in 2002–2003 were doubled in 2006–2007. However, 80 per cent of the funds were allocated for road widening schemes. In 2006–2007, 60 per cent of the funds were earmarked for public transport, which primarily includes the metro system (Ministry of Transport and Power, Government of National Capital Territory of Delhi [GNCTD], 2006). Cars are owned by less than 10 per cent of the households in most Indian cities, except in Delhi. Therefore, an investment in car friendly infrastructure is not meant for the majority of the commuters.

In the name of promoting public transport, demands for rail-based systems – metro, light rail transit (LRT) and monorail – have been made by several cities. This is despite the fact that rail-based systems are capital intensive, and the existing metro systems in Kolkata, Chennai and Delhi are carrying less than 20 per cent of their available capacity. All three systems are running with operating losses. Without questioning the methodologies of demand projections and usage patterns, the government in Delhi has decided to expand the metro system. Similarly the state governments of Maharashtra, Karnatka and Andhra Pradesh have decided to invest in metro systems that will benefit only a small share of journeys.

TRANSPORT POLICY PLANNING

At the policy level, town and country planning organisations and development authorities are expected to prepare their city master plans and city development plans. Since 1960, such plans have been prepared for several metropolitan cities. However, these plans have not been very effective in managing urban growth. Almost all cities have a slum population occupying land that is not earmarked for them.

The growth in any city is almost always accompanied with the expanding size of the urban "informal economy". Larger cities have more slums and squatter settlements. In the million-plus and megacities in India, 40–50 per cent of the population live in informal housing (Misra, 1998). The rising cost of transport within the city combined with long working hours force the workers to live in proximity to their workplace. A large number of people living in these units are employed in the informal sector, providing various services to the outer areas of the city where new developments have been planned. The growth rate of squatter households, as compared to that of the non-squatters, is nearly four times higher in Delhi – we see a 54.2 per cent growth in squatter households compared to a 12.3 per cent in non-squatter households (Hazard Centre, 1999).

Many of the urban population living in informal settlements are captive users of low-cost travel modes (walking and cycling) because many of these residents cannot afford to pay even the low, subsidised fares for buses. For the poorest 28 per cent of the households, with monthly incomes of less than Rs. 2000, a single worker spends 25 per cent or more of their entire monthly income on daily round-trip bus fares. For those with incomes much less than Rs. 2000, the already-low bus fare is prohibitively expensive.

The impact of recent eviction and resettlement policies in Delhi has adversely affected a large number of poor households in the city. People who have been relocated have reduced access to jobs because the new residential locations are 10–15 km away from their previous residences. Hence all walking trips have to be replaced with motorised trips. Often, increased distance from the workplace has also meant increased travel time and expense.

The process of preparing master plans has been in place since 1960. Master plans are made to provide a long-term vision and guidance for the growth of the city. Yet Master Plan Delhi (MPD) 1981 and 2001 failed to provide effective guidance. In 2006, Delhi witnessed demolition of non-compliant uses on a large scale. Why is it that after nearly 50 years of master planning, cities are still struggling to meet the needs of the city residents? A large section of the city population continue to be called "illegal" residents. Is it because of poor plans, or poor implementation, or a lack of resource or expertise?

CAPTIVE USERS

Travel patterns for people living in informal housing or slums are very different from residents in formal housing (Table 1.1). Bicycles and walking account for 66 per cent of commuter trips for those in the informal sector. The formal sector is dependent on buses, cars and two-wheelers.

Table 1.1   Commuting patterns of high- and low-income households in Delhi (1999) and Mumbai (2004)

	High-income households[a]		Low-income households[b]	
	Delhi	Mumbai[c]	Delhi	Mumbai
Cycle	3	3.5	39	6
Bus	36	15	32	14.5
Car	28	3.6	0	0
SC/MC	29	8.5	3	.7
Auto	2	2.1	1	1.3
Rail		21	2	16
Others		1.1	2.	
Walk	2	45	22	61

Notes: [a] Indian Institute of Technology (IIT) survey of high- and middle-income households (average income Rs. 7000 per month). [b] IIT survey of low-income households (average income Rs. 2000 per month). [c] Baker et al. (2004). SC/MC: Scooter cycle/Motorcycle – two-wheeled motorised vehicle, Auto: Autorickshaw – a three-wheeled motorised vehicle.

This implies that despite high risks and a hostile infrastructure, low-cost modes exist because users of these modes do not have any choice – they are captive users.

ROLE OF THE INFORMAL SECTOR AND PARATRANSIT

Urban travel in Indian cities is predominantly walking, cycling and public transport, including intermediate public transport (IPT). The variation in modal shares among these three seems to show a relationship between city size and per capita income. Small- and medium-sized cities (such as Kanpur and Ahmedabad) have a lower per capita income than the megacities. Therefore, the dependence on cycle rickshaws and cycles is greater than it is in larger cities. Delhi is showing declining trends in the three-wheeler population because of restrictions on fuel and the age of public transport vehicles.

Cities with populations of 0.1 million to 0.5 million are characterised by short trip lengths, medium density (~400–800 persons/ha) and mixed land-use patterns. Nearly 50 per cent of the trips are by walking and bicycles, and another 30 per cent are by paratransit. The two-wheeler share ranges from 15 per cent to 40 per cent in some cities and car share remains below 5 per cent (RITES, 1998). Organised public transport services under the public sector exists in few cities; however, several state-run corporations have been found to be financially unviable in the past and have been closed. In some cities, private buses have recently been introduced, but the bus transport operation is predominantly under the public sector. IPT modes such as tempo, auto and cycle rickshaws assume importance as they are necessary to meet travel demands in medium-sized cities in India such as Hubli, Varanasi, Kanpur and Vijayawada. However, there is no policy or project that can improve the operation of paratransit modes. Often, fare policy stipulated by the government is not honoured by the operators, and also the road infrastructure may not include facilities for these modes. As a result, the operators have to violate legal policies to survive in the city.

Public transport is the predominant mode of motorised travel in megacities. Buses carry 20–65 per cent of the total trips excluding walk trips. Despite a significant share of work trips catered to by public transport, the presence and interaction of different types of vehicles create a complex driving environment. Preference for using buses for journeying to work is high by people whose average income is at least 50 per cent more than the average per capita income of the city as a whole (CRRI, 1998). Although an increase in fares may or may not reduce the ridership levels, it will certainly affect the modal preference of a large number of lower-income people who spend 10–20 per cent of their monthly income on transport. A survey result shows that nearly 60 per cent of the respondents found that the minimum cost of journey-to-work trips by public transport (less than Rs. 2 per trip) unacceptable (CRRI, 1998). Even the minimum cost of public transport trips accounts for 20–30 per cent of family income for nearly 50 per cent of the city population living in unauthorised settlements. This section of the population is very sensitive to the slightest variation in the cost of public transport. In outer areas of Delhi, the presence of non-motorised vehicles and pedestrians on some of the important intercity highways with comparatively long trip lengths shows that a large number of people use these modes because they have no other option. Since a subsidised public transport system remains cost prohibitive for a large segment of the population, we should note that the market mechanisms may successfully reduce the level of subsidies; however, they would eliminate certain options for city residents.

The RITES (1998) projected modal shares show that in future bicycles and non-motorised vehicles (NMVs) will continue to carry a large number of trips in cities of all sizes. Public transport trips will be in the range of 25–35 per cent of the total number of trips. Walking trips will constitute 50–60 per cent of total trips. Despite a high share of walk trips and trips by non-motorised modes, the transport infrastructure does not include any facilities for these modes.

ALTERNATE TRANSPORT SOLUTIONS

Traffic and transport improvement proposals prepared by consultants before the JNNURM included proposals for road widening, grade-separated junctions and metro systems. These projects were justified on the basis of forecasts for the next 20 years.

While the road widening and junction improvement schemes were implemented in a few cities, public transport remained in the reports only because the finances required for metro projects were beyond the capacity of state or city governments.

ROAD IMPROVEMENT PROJECTS

Several Indian cities have constructed or made plans for new flyovers. The justification for flyover construction is to reduce long delays at intersections and provide uninterrupted movement for long-distance traffic. However, flyover construction cannot provide long-term solutions because it improves journey times for only a small section of the road for cars that form only 20–25 per cent of the total commuter trips in a city like Delhi. In other cities, car trips are less than 20 per cent. It does not have any benefits for bus commuters because bus stop locations are shifted away from the intersection, increasing the walking distance for changing buses which go in different directions. With an increase in the speed of road vehicles, bus commuters as well as other pedestrians find it difficult to cross the road. Thus, flyovers result in short-term benefits for car users at the cost of increasing traffic hazards and inconveniencing other road users. They also encourage people to use cars and two-wheelers and to move away from public transport, walking and bicycling. This results in more vehicles, congestion and pollution on the roads. A careful look at the road widening and junction improvement schemes shows that widening has been done by reducing the space for pedestrians. Not a single city in India has implemented facilities for bicycles, public transport buses or IPT vehicles like three-wheelers and rickshaws. Junction improvement schemes have included creating free left-turns, shifting bus stops away from the junction and creating grade-separated junctions. In other words, investment in road infrastructure improvement has meant facilities for vehicles that carry a much smaller share of total trips compared to the trips by pedestrians and non-motorised vehicles.

PUBLIC TRANSPORT PROJECTS

Different Indian cities are either implementing or looking at new transport systems, be they metros, high-capacity buses or sky buses. The argument given for introducing new technologies is to serve the high-density demands expected on a few corridors in the city. In the last 15 years, comprehensive traffic and transport plans have been made for at least 15 cities. Travel forecasts for next 30–40 years have been used to justify the proposal of LRT or metro systems. Other than Delhi, no other city has been able to implement these recommendations. Travel demand depends upon city size, trip length, location and density of jobs and residences, and other socio-economic conditions. A high-capacity system requires that the demand for it should exist within walking or easily accessible distance. This requires a very high-density development within the catchment area of the rail system. In Hong Kong and Singapore, high-rise buildings along the metro corridor provide a high density of residences as well as jobs. Therefore, the capacity provided by the metro can be well utilised. Indian cities have high-density developments in the form of urban slums. Even a subsidised metro system is too expensive for slum-dwellers. Therefore, the demand for metro systems in Indian cities is low. This is the reason that the Kolkata, Chennai and Delhi metro systems are carrying less than 20 per cent of the available capacity. The metro and LRT have low social cost in terms of energy consumption and pollution only when the system runs to its capacity. Since the supply exceeds the demand, the system runs at a loss. System demand is dependent on the ease of access, low fares and dependability. The metro is a capital-intensive system (Rs. 200–300 crore/km). It is suitable for meeting the mobility requirements of less than 20 per cent of the city residents because Indian cities have a medium-density development of middle-income groups.

For the same price, 30–50 km of bus network can be developed including the use of modern buses. This would benefit 30–50 times more people than a metro system. The cost of single metro trip is at least Rs. 45 compared to Rs. 15 for a bus trip. Since cars and personal two-wheelers provide a flexible door-to-door service, it is not easy to attract these users to a metro, even if they can afford the cost. Tickets on the metro have to be subsidised at least 10–15 times more than a bus ticket for the same journey.

The efficiency and viability of these systems has to be judged in terms of how well they can serve an individual trip, how many people can benefit for the same investment and how flexible the system is in meeting the changing demands of the city. The metro, LRT and sky bus are all capital-intensive systems and hence cannot match the extensive coverage provided by the road-based systems. All three require a high density of population that can afford the price of rail-based systems living along the corridor. This is not the case in most Indian cities. Therefore rail-based systems have to depend on buses, three-wheelers and rickshaws as feeder modes to increase their catchment area. A large number of people should be able to access the metro without depending on a feeder mode. Only long-distance travellers (trip length at least 15 km) are likely to use a feeder mode. Therefore, in order to realise the social benefits of metro systems the city structure has to change completely. The advantage of a road-based mass transport system is that, at a cost of one-tenth to one-twentieth of that of other mass transport systems, a high-quality public transport system can be provided within walking distance. Since the catchment area of a road system depends on the extent of the road network, road-based systems are capable of reaching almost 80 per cent of the city population. Access to this system includes improved and safe pedestrian paths. If this is done, this system will be able to match the convenience and flexibility provided by private modes. Consequently, we can expect a high demand and, therefore, better capacity utilisation of this system. One of the major drawbacks of rail-based systems all over the world, including in Kolkata, Chennai and Delhi, has been their low-capacity utilisation. Even if better feeder trip systems are planned, capacity utilisation is not expected to change because only long trips (longer than 15 km) will benefit from these systems. Bogota and Curitiba, both considered model public transport systems, have decided to expand their bus rapid transport (BRT) system in order to serve the whole city with public transport that does not require subsidy, that restricts car usage and has a major impact on safety, pollution and energy consumption.

WHAT IS THE IMPACT OF NUTP AND JNNURM?

Under JNNURM, the government of India has identified 63 cities for the provision of assistance to upgrade the infrastructure. Detailed guidelines have been provided to ensure that public transport gets priority in these cities. For getting approval for transport projects, the guidelines recommend that the transport infrastructure improvement schemes have to be in compliance with the National Urban Transport Policy (NUTP). Since the focus of the NUTP is public transport, pedestrians and bicycles, cities are modifying the earlier road expansion projects to BRT- and bicycle-inclusive plans. BRT- and bicycle-inclusive plans for five cities have been approved by the central government and another five cities are at different stages of preparation. However, pedestrian and bicycle facilities are not the focus of these projects. In a six-lane arterial road, two lanes are reserved for public transport buses; however, there is a reluctance to provide quality facilities for pedestrians and bicyclists. This is reflected in the priority for space allocation for various modes in a restricted right of way. In order to accommodate two lanes for cars and an exclusive lane for buses, pedestrians and bicyclists have been given less space than desirable. This is despite the fact that nearly 50 per cent of trips are either by pedestrians, bicycles or IPT systems. The main motivation for preparing BRT projects has been to become eligible for the grant aid offered by the central government at the earliest opportunity. It is yet to be seen whether public transport, NMV and pedestrian friendly infrastructure is created when these projects are implemented.

The implementation of the BRTS has commenced in Delhi. However at times it seems that accommodating the demands of the major stakeholders of the "transport industry" – the Delhi Metro Rail Corporation (DMRC), the public works department, the light rail and monorail industry – in the planning and investment agenda is the primary focus. The first phase of the metro is carrying 20 per cent of the projected trips and facing operating cost losses, yet extensions of the metro line are being actively pursued by the government – both by the bureaucracy and the politicians. Providing an efficient and safe transport for the masses, and using public money in the most efficient way, is not the driving force for implementing the BRTS in Delhi. The company that has been appointed to implement the project – Delhi Integrated Multimodal Transport System (DIMTS) – is also preparing plans for light rail transit and monorail. The BRTS road designs have been modified to "improve" car flow so that after the construction of the BRTS lanes, car users are not inconvenienced, even if it means reduced level of service for pedestrian and bicycle facilities.

TRENDS IN OTHER REGIONS

The rapidly expanding cities in other parts of the world (such as Shanghai, Mexico City and Johannesburg) are characterized by a significant proportion of the city population being dependent on the informal sector. A majority of the population is dependent on walking, bicycling and public transport. Automobile mobility is still the preserve of the minority.

Historically these three expanding cites – Shanghai, Mexico City and Johannesburg – have been different in many aspects. Shanghai, until the mid-1980s, had a controlled economy and invested in public housing and bicycle infrastructure. A large number of people employed in the informal sector were counted in the floating population. However, since the opening of the economy, major changes were brought about in government policy and the car industry was declared a pillar industry in China; in the last decade, elevated highways, satellite towns and monofunctional districts have put human-scale transport infrastructure on the back burner. Shanghai's official policy is to reduce cycling, which led to a drop from almost 40 per cent to 25 per cent of all trips between 1995 and 2004. The city is successful in attracting greater car use, which doubled during the same period, leading to increased average commuting distances of up to 70 per cent. In Mexico City, hundreds of thousands of houses built in outlying districts like Ciudad Neza were considered *asentamientos irregulars* or *asentamientos paracaidistas*. However, since the official policies have supported them and investments in upgrading the infrastructure have been made, the housing colony that started as traditional squatter settlements has been transformed into a vibrant city of 1.5 million people. A large number of businesses have been set up within the settlement, providing nearly 65 per cent of jobs for the residents. Despite this, the current policies have been focusing on creating a Santa Fe-like development, which is designed as a multifunctional cityscape, including corporate towers and commercial facilities characterised by a highly exclusionary character. Around 50,000 minibuses and microbuses handle a majority of the trips in Mexico City; however 40 per cent of the city's transport budget between 2000 and 2006 was spent on its Segundo Piso, an elevated highway used by less than 1 per cent of its residents. Johannesburg has a history of apartheid. Like most other cities of the global south, a large part of the local economy is generally thought of as informal. The informal sector is growing very rapidly. It accounts for a sixth of all jobs. The apartheid legacy manifests itself in the contrasts between Soweto and Johannesburg, both of which are integral parts of the same city. Like many other apartheid cities, Johannesburg also has a sprawling reality with density levels that are too low for public transport. At the same time there is high population density in townships like Soweto and other informal settlements that have mushroomed on the outskirts of the city. In the apartheid era, these were deliberately surrounded by high-speed freeways and rail lines, acting like moats to keep people in rather than out. Johannesburg's public space has been taken over by traffic, as is shockingly illustrated by its annual accident statistics of 56 fatalities per 100,000 inhabitants compared to 3 in London and 7 in Mexico City. The city seems to have surrendered to the safe, private environments of shopping malls. Marginalisation and containment, planned under apartheid, has been perpetuated in the post-apartheid period. The percentage of stranded people who walk to work (often in dangerous circumstances) for more than 30 minutes, because they cannot afford any form of public transport, has increased. About 46 per cent of households are spending more than 10 per cent of their income on transport. Minibus taxis, the major public transport mode, receive no operating subsidy. But the provincial government is planning to invest US $2.7 billion for a rapid rail project.

It is clear that in the name of development and progress, auto-based mobility solutions have dominated the public transport policy agenda and investment in these rapidly expanding cities. Sustainable transport concepts are used more to promote capital-intensive systems like heavy rail, which may not have an extensive catchment area but require huge capital. Modes of transport that are used by a majority of the people, such as walking, cycling and microbuses, do not become the focus of a sustainable transport policy agenda. At the same time the mature cities that developed around rail-based public transport networks, ended up becoming dependent on automobile-based mobility. In the last decade there have been serious efforts towards a more sustainable approach, which includes promotion of bicycling, bus transport and pedestrianisation. However, despite investments and expertise, the process of moving towards public transport and non-motorized mobility has been rather slow.

CURRENT KNOWLEDGE BASE

The current understanding of sustainable transport policies recommends mixed land-use patterns and high densities in cities. This ensures short trip lengths suitable for walking and bicycling trips. In comparison to cities in the West, cities in low-income countries consume less transport energy. High densities, intensely mixed land use, short trip distances and a high share of walking and non-motorised transport characterise these urban centres. However, their transport and land-use patterns are so confounded by the spectre of poverty and high levels of complexity that it becomes difficult to analyse their characteristics using the same indices as used for cities in highly motorised countries (HMCs).

Land-use policy can influence city density, structure, diversity and local design to influence urban air pollution and promote sustainable transport (Gwilliam, Kojima and Johnson, 2004). In response to these recommendations, cities plan high-rise structures, commercial centres close to metro stops, and pedestrian and bicycle facilities in residential areas. However, a high-density population existing in slums and street vendors near the bus stops and along the roads are not considered desirable for sustainability. Three-wheelers and rickshaws are believed to create congestion. The current knowledge of city planning processes and transport systems does not have solutions for the existing complexities.

The idea of sustainability in low-income cities and mature Western cities may well be different. In Delhi, environmental concerns end up in measures to "clean" the city of pollution and traffic congestion. Both of these at times hurt the poor by evicting them from the city centres and industrial jobs or by increasing the costs of transport. There is a conflict in the meaning of sustainability for the rich and the poor. For the rich, sustainability can be achieved through technological fixes – cleaner engines, fuels and metro systems. For the poor, sustainability is a lifestyle and employment issue better addressed by city form, land use and accessibility issues. This conflict manifests itself politically, with no easy or quick solutions.

ROAD TO FUTURE TRANSPORT

The growing cities, mostly in Asia, Africa and Latin America, are characterised by diversity and heterogeneity in socioeconomic conditions. These megacities are agglomerations of several small cities having multiple economies, in close proximity to each other. One economy serves the needs of the affluent and features modern technologies, formal markets and the outward appearance of developed countries. The other serves disadvantaged groups and is marked by traditional technologies, informal markets and moderate to severe levels of economic and political deprivation. A majority of the population is dependent on walking, bicycling and public transport. Automobile mobility is still in the preserve of the minority. The current understanding of transportation issues in these cities has prompted "improvements" in the transport situation by breaking up public spaces for the uninterrupted movement of private vehicles. Improvement in road capacity in these cities has meant reducing pedestrian and bicycle facilities, removing street vendors, restricting pedestrian movements and constructing grade-separated junctions.

If sustainable transport is to be promoted, then the following must be kept in mind:

1. Stricter emission laws in the West did not reduce total emissions in the West.
2. At present, many low-income countries have a desirable modal share. The challenge is to preserve this in the future amid growing car and two-wheeler ownership rates.
3. If the current shares of walking, bicycling and use of public transport have to be preserved and encouraged, then clean technology alone is not sufficient. Implementation of safe infrastructure for walking, bicycling and public transport will have to be prioritised.
4. Sustainable transport needs inclusive cities, and inclusive cities are also safe cities. Sustainable transport takes us beyond physical infrastructure. It also provides the opportunity for honest living. Inclusive streets ensure not only safe mobility, with reduced risks of traffic

accidents, but also reduced street crime and better social cohesion, making public transport attractive and the preferred choice for commuting.

Over the last decade, there have been serious efforts in several cities to put land-use and transport strategies closer together. However, despite investments and expertise, the process of moving towards more sustainable urban structures where movement is based on public transport and non-motorised mobility has been rather slow. If cities in the future will have to rely on sustainable transport, then we have to move fast towards understanding the forces that promote car use.

REFERENCES

Baker, J., Basu, R., Cropper, M., Lall, S. and Takeuchi, A., 2004. *Urban Poverty and Transport: The Case of Mumbai* (personal communication).

CRRI, 1998. *Mobility Levels and Transport Problems of Various Population Groups*. Delhi: Central Road Research Institute.

Gwilliam, K.M., Kojima, M. and Johnson, T., 2004. *Reducing Air Pollution from Urban Transport*. Washington, DC: World Bank.

Hazard Center, 1999. *This City Is Ours: Delhi's Master Plan and Delhi's People*. Delhi: Sajha Manch.

Ministry of Urban Development, 2006. National Urban Transport Policy, Ministry of Urban Development, Government of India.

Misra, R.P., 1998. Urban India: Historical Roots and Contemporary Scenario. In *Million Cities of India*, R.P. Misra and K. Misra (eds), p. 52. Delhi: Sustainable Development Foundation.

RITES, 1998. Traffic and Transportation Policies and Strategies in Urban Areas in India, Final report prepared for the Ministry of Urban Affairs and Employment, Government of India.

Chapter 2

From myth to science in urban and transport planning

From uncontrolled to controlled and responsible urban development in transport planning

Hermann Knoflacher

CONTENTS

INTRODUCTION

It sounds astonishing to talk about myths in a world dominated by natural science and rationality, and it is even more astonishing that this should happen in the transport-engineering and urban planning world. In human history myths were always appearing when society was not able to understand or explain phenomena in the surrounding environment. For thousands of years these myths were related to natural phenomena from our universe and our natural environment. When natural science began to discover and explain the mechanisms behind these visible phenomena, these kind of myths disappeared.

One of the dreams of mankind – that of effortless, convenient and fast physical transport – became reality when the steam engine, and later the combustion engine, came into use for transport. The effects of these wonderful new things have not been understood till recently, especially by those responsible for planning, building and operating these inventions from the 19th century, and we need to take account of the effects these changes will produce. Things in this technically dominated field happen outside the perceptional borders of our senses, without direct feedback to our actions. The causal loop between cause and effect is cut, and this is the breeding ground for the myths that now exist in many kinds of technology as well as being the breeding ground for ideas and principles in the prevailing transport engineering and urban planning societies.

THE SOCIAL SYSTEM AND THE URBAN STRUCTURE

Settlements and cities have not just happened or and were not just "founded" by somebody, as is written in many textbooks for urban planning. The precondition for any built settlement is a society that has developed a level of social cohesion for common sense and a developed social structure and social behavior. *The city is therefore the built expression of the social structure of the society that created the city.*

Historical urban structures are the physical containers of knowledge for over 10,000 years of social development toward the creation of a sustainable society.

In this way many societies failed if they made mistakes by overexploiting the resources for urban life and/or the human resources of the city population. These cities and their societies have disappeared over time. The social system in an urban structure means that, over the whole history of urban life, people have devoted a certain amount of their personal energy toward the construction and maintenance of the urban system they lived in. Only a few, the rich or the powerful, were able to separate themselves from society or from the social network of the urban population, as long as this society accepted their extraordinary position or was supportive of it.

For millennia, the body energy of pedestrians, the main mode of ubran mobility, was limited and therefore cities could only be sustainable when taking into account these limitations by creating a well-balanced structure (trade-off) between the private and public spaces of and for their citizens. All sustainable urban structures have, therefore, a finely tuned public space, which we call the *urban street network*, created on a human scale with a network width (distance between streets or pedestrian passages between or through the buildings) of about 20–50 meters. This life-supporting urban street network can be found in all the historical urban structures of cities around the world. Modern transport science can now prove the necessity of this kind of structure since the relationship between the acceptability of walking distance and body energy consumption, a "law of human behavior," was also discovered in these historical structures (Knoflacher 1981a,b).

FASCINATION OF ARTIFICIALLY DRIVEN TRANSPORT MODES – THE VEHICLE AND THE TRANSPORT INFRASTRUCTURE

When the steam engine was connected with wheels and put on rails to enable movement the situation in the physical patterns of built environments began to change. When rails were introduced the main effects were the ability to transport large quantities of goods at high speeds independently from local energy sources and to move masses of people between distant places. The effect on the cities was not so great, since the local transport systems were still dominated by pedestrians, horses and ox carts. In the second half of the 19th century, public transport was introduced in larger cities in the form of street cars, traveling at a speed four to five times that of pedestrians. The city size could now exceed the former limit of about 800,000 people to 1 million to 10 million, or even more. Some societies, such as those in China, were clever enough to cycle, also an invention of the 19th century, as a kind of sustainable mass transport mode, one of the cleverest mechanical modes ever invented. But still human body's energy had to be used. Slightly older than the cycle (draisine or velociped [1817–1819]) is the car (the 1784 steam carriage of W. Murdoch and the 1801 full-sized road vehicle designed by Richard Trevithick). The car has replaced human body energy in creating movement and putting the driver in a convenient sitting position, giving him the expression that he can "fly over the ground", conveniently and effortlessly. Modern science knows that car drivers use less than one-third of the body energy per minute compared to pedestrians, but can move 10 or 20 times faster. For human perception, this is really a miracle; for traditional transport system planners and politicians, it is the base of the myth that travel time is saved in urban transport systems.

THE MIRACLE OF EFFORTLESS, INDIVIDUAL MOBILITY – THE CAR AND ITS EFFECT ON THE URBAN SYSTEM

Since there was some public space available, this new, effortless, individual mode of the car had most of the advantages and opportunities compared with existing transport modes from the outset. This development was supported by "experts" and used by institutions like technical universities, where students could learn about the physics of these new mechanical modes. Transport engineers were trained to provide the infrastructure, the operation and the maintenance of this wonderful system in principle to fulfill what was called the "travel demand" of the industry and later what they called "society". Speed needs space due to the limitations of human senses, which were not developed for reactions at high speeds but rather for pedestrian speeds, in terms of reacting in a responsible manner in time and space. It is very important to distinguish between public transport and private cars. Public transport is a so-called closed system, with controlled access at certain stopping points and clear transportation conditions and responsibilities. Car traffic is a so-called open system – as it has been developed so far – with uncontrolled access and without a clear responsibility. The responsibility split between system users, administrations, land owners, etc. had already had a certain effect on the urban structure as has been explained before. But compared to the effects of cars on the urban structure, that effect was of minor importance. Cars are "eating up" public space

(a) by parking
(b) for movement.

When vehicles are moving the space needed increases by the square of the speed due to the braking distance. Since public space in a human-scale city is limited, car traffic became what we currently call congestion. The multifunctionality of the urban road system was given up for the benefit of the free movement of cars. The precious and highly valued public space was downgraded for two purposes, driving with speed and parking, without any compensation to society. Traffic engineers were focused toward mechanical transport modes but nothing else, an approach that was very narrow minded. Architects were educated to build houses without any taking account of the complex social structure of the urban framework. These two disciplines began to transform cities everywhere, but with no idea and no education about the effects these activities would produce on the environment, traffic safety and society. To have well-meaning intentions without knowing the real system effects is a terrible kind of irresponsibility, and this is the general situation today.

TRAFFIC ENGINEERING – AMERICANISM IN THE TRANSPORT SYSTEM FROM A COUNTRY WITHOUT AN URBAN HISTORY

America was the continent in which European inventions in the transport sector became an industrial product: the car. Since Native American society was all but extinguished the country was thinly populated and had no historical urban structures. To control these open spaces the car was ideal. In Europe, where the use of motorized transport started at the same time as in the United States, the process was stopped by the Second World War. After the end of the war, car traffic began to influence the settlements in the United States in a totally different manner compared to the whole human history: new settlements were called cities, but they were not cities at all;

they were what, in Europe, would be termed "urban sprawl". Space had no value any more. There was plenty of space available to be paved with motorways and parking lots.

Society was not important in this world of cheap energy flow; it was much more about what we call individualism or egoism in a race for personal wealth, which is still the dominating paradigm today, not only in the United States but also in other systems, which cannot be called societies any more. The construction of motorways for free movement of cars was the main focus for traffic engineers and urban planners. Everything else had to be subordinated. American urban transport planners had till today no idea what a real city at a human scale would be like. Even famous sociologists like Jane Jacobs are not able to express and describe a real human-scale city, because they are totally captured by the presence of cars in the urban space. This kind of American traffic engineering became the dominant paradigm for transport and urban planning in most of the other countries of the globe, where engineers and urban planners applied this narrow-minded principle in an uncritical way. In many of the countries the precious urban structure from the past was ruthless damaged for the benefits of so-called free car flows.

INDIVIDUAL EXPERIENCE AND SYSTEM EFFECTS

It seems obvious that this wonderful mechanical mode – the car – enhanced what we call "mobility", because the term mobility was restricted to car driving. People's personal experiences showed everybody that driving a car "saves time" because you can drive from A to B in a shorter time compared with other modes. Time-saving by increasing speed was the most important basic indicator for the calculations of benefits for the investments in fast transport systems. The real system behavior was not understood so far by most of the experts working in this field. For about 30 years the body of researchers and practitioners have understood that, in the transport system, no time-savings can occur if speeds are increased. It is not only that the travel time budget remains constant even if the speed of the system increases, but also that the travel time distribution among different transport users with different speeds is the same. There is enough empirical evidence that this has been happening everywhere around the globe and there is no empirical evidence to contradict these facts. Scientifically based transport and urban planners therefore deal in a quite different manner with speed. If speed is increasing, then the distances are increasing, and this is changing the urban structures. Multifunctionality and variety is only possible and sustainable in a transport system with low speeds. If we increase the speed the sustainable urban structure is destroyed and replaced by urban sprawl – mainly for housing and the concentration of economic activities in places chosen by industry and not controlled any more by the society. Fast transport systems make urban administrations and even country administrations helpless. High speeds benefit big corporations and centralized financial powers. High-speed individuals – if they are rich enough – as well as corporations or big companies get the opportunity for a free choice of location. However, society and their elected bodies are losing their power, because that power is limited at the administrative borders. The new plutocratism, whether private or corporate, is now able to play the different interested parties against each other. This is happening now on the community level, the country level and the state level as well as on the international level. There is only one group of winners: international corporations without space limitations. The loser is everybody else. When sprawl is the built environment of a socially degraded society there is a step backward from more than 10,000 years of human development. It is the result of the extrapolation of individual experience on the system. They are not the system, but the constituents of the system with new emerging attributes, never experienced before. The calculations are based on illicit extrapolations of personal experiences and plausibility. They calculate benefits from travel time savings in the system in which reduction of travel time by enhancement of speed doesn't exist. It is much more the behavior of a guild than one of science.

URBAN AND TRANSPORT PLANNING AND RESPONSIBILITY FOR SYSTEM BEHAVIOR – THE PARADIGM CHANGE

Science based on theory and empirical evidence has provided enough knowledge during the last 30 years to prevent the mistakes in these fields from the last 150 years from happening again. Following basic system behavior facts have to be taken into account:

There is no growth of mobility in the system if we define mobility in a scientifically sound manner. Mobility can only be defined if it is purpose-related. Nobody leaves home without any purpose. The cause of each trip is the unmet need at home, e.g., work, shopping, etc. The number of purposes has not changed with the changing transport environment. People are going to work, school, shopping, for social contacts, leisure and other purposes just as they did before the transport system became mechanized, so the number of trips is the same. If car trips are increasing, all other kinds of trips are decreasing by the same amount. The total number of trips therefore remains the same. What we can change is the *kind* of mobility, but not the mobility itself. A person who is talking about increasing mobility is talking about an unrealistic system.

The second fact is the *constancy of travel time*. Transport speed cannot save time in the system, it can only change the distances. A good urban structure can fulfill all the needs of society with a minimum of distances – this is the basis of cities. The duty of urban transport planners is therefore to minimize trip lengths by creating a complex multifunctional urban structure with a minimum amount of public space for movements and a maximum of functionality in this public space. This is only possible if low speeds dominate. Allowing cars to park in public spaces is wasteful of public space and money. So no parking can be allowed on streets and roads, only efficient public transport should be allowed to use public space, with a higher speed, if it is organized in an efficient way.

Energy consumption increases with increasing speed, and this makes cities and societies dependent on energy imports, which will become more and more expensive and finally, within a visible time period, impossible. Urban structures are long-living; their time span is longer than that of cheap fossil energy. Only irresponsible planners and politicians can therefore provide an expensive and urban-damaging transport infrastructure without any long-lasting future. Finally, there is a third wrong assumption in traditional planning:

There is no freedom of modal choice. "Modal choice" is a technical term, used in transport models to describe the choices individual or groups to select transport modes for their particular trips, expressed by the share of trips made by mode. Scientific research has proved that people are "car-addicted". Since the car replaces physical body energy, it gets in touch with the human body at its energetic level, which is the deepest-rooted level of all beings. The only way to make people free from car addiction is therefore a structure that prevents this addiction. This means that a physical distance between different human activities, such as living, working, shopping, etc., and their parked cars must be at least as long as the physical distance to the next public transport stop. This means a responsible urban planner and politician has to change the existing parking regulations fundamentally, introducing an equidistance between parked cars and public transport stops as well as introducing the principles of a market economy in the transport sector. Those who are using more public space have to pay more and those who are using the privilege of parking at home also have to pay more. To compensate for the benefits of parking at home and at destinations, a levy must be implemented, in accordance with the distance between parking place and origins or destinations. Increasing in distance, beginning with 200m, the levy can be reduced with the decreasing irritation of car use. If the money is used to reshape the public space and the urban structure in favor of modes agreeable with a sustainable city, pedestrian, public transport and cycling, we can create a push and pull effect to help people to escape from the car-trap. There is no need for discussion or debate and no need to hesitate.

MASTERPLAN, "TRAFFIC" FORECAST – THEORY AND PRACTICE

The purpose of an urban masterplan is the controllable, fruitful and, in some cases, sustainable development of a city – in theory. The practice is quite different. Everywhere in the world, independently from the political system, urban development is not following the masterplan, not even its principles. Agricultural land and farmhouses are converted into apartments with swimming pools for rich land owners, shops are opening in places where they should not be and what was planned by land-use planners is neglected by developers. On the other hand, transport planners are making "traffic" forecasts for car traffic in a totally "unconditional" way (without any conditions), but with a certain degree of accuracy. If the accuracy is high they believe that they have made a good scientific forecast. But compared to their promises that they will solve transport problems, they have not only been unable to solve transport problems within the car traffic system, but they have also created many other problems, such as deficits in public transport, traffic accidents, air pollution and unbearable noise levels in the urban environment.

Was this the failure of the professional experts or is it the failure of the decision makers, the politicians? In general decision makers are implementing what experts are telling them is needed, but if the system doesn't work as expected, the experts tend to blame the decision makers for their decisions.

If we look at the problem in a systematic way, we see, on the one hand, the city and the traffic system and, on the other hand, the system of people, experts and decision makers, in the widest sense (administrations, politicians and lobbies). One part of the system wants to regulate the other: the experts and decision makers want to regulate the city and the traffic system. We have to look at the degree of freedom in the different systems.

If two systems are in interaction and have different degrees of freedom, the system with the greater degree of freedom will always control the system with the lesser degree of freedom.

This principle – a basic principle of evolution – can be expressed in the following form: "The variety of a regulator must be equal to or greater than the variety in the system being regulated", or "The greater the variety within a system, the greater its ability to reduce variety in its environment through regulation". These principles are known as the "law of requisite variety".

In practice it is therefore obvious that the urban system itself is the regulator and not the experts or the decision makers. In the transport system, traditional educated experts are clearly slaves or servants of the regulator "Eigendynamik of the car traffic". It is obvious that traditionally educated experts have not understood the system principles in such a way that they can regulate the development of the city or even the transport system. They are treating symptoms, or trying to control the unregulated system, using laws depending on decisions of the courts, a very costly and, to a certain degree, also unfair measure.

The reason both disciplines were unable to give the right advice to the decision makers was their ignorance about the system behavior that had changed with the mechanical transport modes, especially the car. If individuals or companies have immediate access to a fast, cheap (paid for by somebody else) mechanical transport system, they get a degree of freedom that is always much greater than the degree of freedom of their counterparts: planners, administrations and politicians. They can choose the places of living, working or leisure in- or outside the administrative borders of the city. They are optimizing their individual positions at the expense of the whole of society and the future of the city. If a city tries to stop this strategy, they can even cross over the administrative border and settle in a neighborhood community, a strategy used by international corporations to get the best conditions for their branches by force.

Direct proximity between car and man, or a company with the motorized individual mode, that enhances the degree of freedom of individuals or companies, far away from the regulatory power of the city administrations and their politicians.

If city administrations and politicians are really interested in sustainable development of their society, they must regain a greater degree of freedom than the single elements. This is only possible if they make the parking regulations, separating parking from human activities. If they are not able to implement and monitor these principles they are restricted to only treating symptoms, which they can do as much as they want, but without any success, whether for urban development or traffic solutions. They will be controlled by the "Eigendynamik", which is much stronger than any regulatory power, since it is driven by the deepest levels of human evolution.

Today the development is characterized by a self-driven dynamic of the motorized transport system, which is reducing variety in its environment through regulation. The worst examples are elevated motorways, flyovers or motorway intersections built in the living urban body.

THE CHALLENGE OF THE PARADIGM CHANGE FOR PRACTITIONERS AND DECISION MAKERS: HOW TO ESCAPE FROM THE PATH TO AN UNSUSTAINABLE AND DESTRUCTIVE URBAN FUTURE?

This is a really crucial question. In a society of addicts it is difficult to bring about structures to "heal" the urban transport. Healing the people from "car addiction" also heals the cities. There is enough scientific evidence from behavioral science, from traffic engineering, from urban analysis and from economic analysis to indicate that this is a necessary change in order to continue human development in the urban context. No matter how difficult, it is the responsibility of people who have recognized how dangerous this part of development is– which was extremely convenient but at the same time extremely damaging for the urban structure – and that it has to be changed. But there are many examples, mainly in Europe, that show that it is possible to escape from this trap, even in highly motorized societies: some cities have introduced structures for the benefit of pedestrians and cyclists as well as giving clear priority for public transport before cars. The pressure on cars is increasing continuously in all cities struggling for survival. It should be much easier in countries where the degree of motorization is still low to prevent society from falling into the same traps that exist for this extremely costly and dangerous development. India, with its old urban culture and its wonderful urban structures, has more opportunities than even European countries to provide a much brighter future for its societies than many other places around the globe. If practitioners and politicians are clever enough to learn from the mistakes that happened elsewhere and can organize their planning based on sound scientific findings instead of importing misleading technical myths, they can open the door for a sustainable future for Indian cities. In doing so they will have to change transport planning rules and stop the construction of urban-damaging transport infrastructures The key question is whether they are able to control the structure of parking. If we are not strong enough, not wise enough and not powerful enough to get parking under the control of society and the administration, they will lose the battle and the cities will follow the same path of destruction.

POLICY GUIDELINES AND AREAS FOR FUTURE RESEARCH

1. Qualification for advisers, consultants and experts has to be checked.
   Due to the far-reaching effects of structural changes in the transport system on the economy, the social systems and the environment, advisers and experts have to understand the system behavior they are dealing with. If they believe the traditional myths, such as "growth of mobility", "time-saving by increasing speed" or "freedom of modal choice", they are not qualified to treat technical transport systems in a responsible manner at the level of transport and urban planning. They are qualified for the lower level of projects, but not the decision level.
2. Transport and urban planning can only be carried out on the basis of external, superficial goals with clear (and, as far as possible, quantitative) indicators.
3. Transport and structural urban planning cannot be separated from each other.
4. *It has to be clarified whether urban and transport planning is done for people or machines (car users).*

5. Without taking into account real physical human behavior, transport planning and decision making cannot be done in a responsible way.

6. Cities for people need a clear policy structure with clear priorities:
   a) pedestrians
   b) two-wheelers, not motorized
   c) public transport
   d) car traffic only for special purposes

7. The physical, financial and organizational structure must represent the hierarchy of the previous point.

8. Cities which need the minimum amount of external energy will survive the future energy crisis. Therefore clear priority is necessary for:
   a) pedestrianization
   b) cycles
   c) public transport before individual mechanical driven modes

9. Trying to solve car traffic problems by treating traffic flows wastes time energy and money. The most efficient and effective measure is the reorganization of parking, as has been proved, so far, wherever it has been tried.

10. Using policy guidelines based on sound scientific principles and not on assumptions.

11. Areas for future research are coming from the open questions. Examples are:
   a) The variety and density of human opportunities and modal split is an interdisciplinary research area that can be carried out at the national level as well as at the international level.
   b) The importance of pedestrians and two-wheelers for a sustainable city.
   c) The optimal integration of surface public transport into the urban structure.
   d) The cost-benefit analysis of traditional car-oriented solutions compared with scientifically based transport and urban planning solutions.
   e) How to reorganize Indian cities toward a sustainable future in the most efficient way.
   f) The long-term economic competitiveness of cities and the transport system.

BIBLIOGRAPHY

Knoflacher, H. (1981a): Zur Frage des Modal Split. *Straßenverkehrstechnik*, (25)5: 150–154.

Knoflacher, H. (1981b): Human energy expenditure in different modes: implications for town planning. In: International Symposium on Surface Transportation System Performance. US Department of Transportation, Washington DC.

Knoflacher, H. und Gatterer, W. (1986): Untersuchung des spezifischen Energieverbrauchs einzelner Verkehrsteilnehmer. Forschungsarbeit im Auftrag des Fonds zur Förderung der wissenschaftlichen Forschung und des Bundesministeriums für Bauten und Technik, Wien. unpublished.

Knoflacher, H. (1995): *Fußgeher- und Fahrradverkehr: Planungsprinzipien.* Böhlau Verlag, Wien – Köln – Weimar.

Knoflacher, H. (1996): *Zur Harmonie von Stadt und Verkehr. Freiheit vom Zwang zum Autofahren. 2., verbesserte und erweiterte Auflage.* Böhlau Verlag, Wien – Köln – Weimar.

Knoflacher, H. (2001): *Stehzeuge. Der Stau ist kein Verkehrsproblem.* Böhlau Verlag, Wien – Köln – Weimar.

Knoflacher, H. (2003): The role of modal split. In: 16th International Symposium on Theory and Practice in Transport Economics – 50 years of transport research: Experience gained and major challenges ahead (Topic 3: Sustainability of Transport: The Role of Modal Split and Pricing), European Conference of Ministers of Transport (ECMT), Budapest, 29–31 October 2003.

Knoflacher, H. (2007): *Grundlagen der Verkehrs- und Siedlungsplanung: Verkehrsplanung.* Böhlau Verlag, Wien – Köln – Weimar.

Chapter 3

The neo-liberal urban development paradigm and transport-related civil society responses in Karachi, Pakistan

Arif Hasan

CONTENTS

INTRODUCTION

The modernist paradigm of urban development, though not officially discarded, has had the culture and economics of globalisation superimposed upon it. New terms (such as Build-Operate-Transfer (BOT), DFI, world-class cities, investment-friendly infrastructure, it is not the business of the state to do business, privatisation, structural adjustment) and the processes linked to them, along with new actors in the form of international capital, are increasingly determining the shape and form of our cities, especially relating to the housing and transport sectors. The new paradigm is capital-intensive and has little respect for environmental and social issues. Based on a study by the Asian Coalition for Housing Rights (ACHR), with which the author was involved, this chapter explores the effect of the new paradigm on Asian cities. It then focuses on Karachi, specifically on the work of the Urban Resource Centre (URC) in trying to create a space for consensus building between various actors in the urban drama, so that communities and weaker interest groups can regain ownership of the city. Much of the URC's work has been related to transport sector projects and processes.

ACHR'S UNDERSTANDING ASIAN CITIES PROJECT

The ACHR is an Asia-Pacific network of professionals, non-governmental organisations (NGOs) and community organisations. Its headquarters are in Bangkok. The decision to create the ACHR was taken in 1987 and this was formalised in 1989. Its founding members were professionals and NGO and community projects working on housing and urban issues related to poor communities. Since then, through an orientation and exchange programme between innovative projects and interested communities and professionals, the network has expanded throughout south, southeast and east Asia. Links have also been created with central Asia and Africa through the savings, credit and housing programmes of SDI (formerly known as Shack Dwellers International).

The ACHR senior members have been very conscious that conditions at the local and international levels today are very different from what they were in 1989 when the ACHR was created. They are also conscious that these conditions are affecting the shape and form of our urban settlements and the living conditions of the poorer sections of society. As a result of this consciousness, in 2003 the ACHR decided to carry out research on a number of Asian cities in order to identify the process of socio-economic, physical and institutional change that has taken place since the ACHR was founded; the actors involved in this change; and the effects of this change on disadvantaged communities and interest groups. Eight Asian cities, along with researchers, were identified for this purpose. The names of the researchers and the titles of their reports from which this chapter is derived are given in Appendix 1; and the objectives of the research and the terms of reference for it are given in Appendix 2. Not all researchers followed the terms of reference strictly. However, an enormous amount of material regarding these cities has been generated and is available with the ACHR Secretariat. The research and logistics related to it have been funded by German funding agency, Misereor.

It was decided that the cities chosen for the research should be as different from each other as possible in political, social and physical terms and that all the researchers should be local people. A synthesis of the case studies has been prepared by David Satterthwaite, a senior fellow of the International Institute for Environment and Development (IIED), UK, and published by the ACHR under the title "Understanding Asian Cities".

The cities chosen for the research were: Beijing (China); Pune (India); Chiang Mai (Thailand); Phnom Penh (Cambodia); Karachi (Pakistan); Muntinlupa (Manila, Philippines); Hanoi (Vietnam); and Surabaya (Indonesia).

FINDINGS OF THE RESEARCH

The research has identified many differences between the eight cities. However, there are a number of strong similarities which are the result not only of how these cities have evolved historically but also of the major changes that have taken place in the world since the late 1980s. It has been identified that these changes are the result of structural adjustment, the World Trade Organization (WTO) regime and the dominance of the culture and institutions of globalisation in development policies (or lack of them) at the national level.

The most important finding of the report is that "urban development in Asia is largely driven by the concentration of local, national and increasingly, international profit-seeking enterprises in and around particular urban centres"; and that "cities may concentrate wealth both in terms of new investment and of high-income residents but there is no automatic process by which this contributes to the costs of needed infrastructure and services".[1]

Many of the other findings of the report have a direct bearing on transport-related and governance issues. A synopsis of these findings is given below.

Globalisation has led to direct foreign investment in Asian cities along with the development of a more aggressive business sector at the national level. This has resulted in the establishment of corporate sector industries, increased tourism, building of elite townships with foreign investment, gentrification of the historic core of many cities and a rapid increase in the middle classes. Consequently, there is a demand for strategically located land for industrial, commercial, tourist and middle-class residential purposes. As a result, poor communities are being evicted from land that they occupy in or near the city centres, often without compensation, or are being relocated formally or informally to land on the city fringes far away from their places of work, education and recreation and from better health facilities.[2] This process has also meant an increase in land prices, meaning that

---

[1] David Satterthwaite; *Understanding Asian Cities. A synthesis of the findings from the city case studies*, ACHR, Bangkok, October 2005.
[2] ACHR monitoring of evictions in seven Asian countries (Bangladesh, China, India, Indonesia, Japan, Malaysia, Philippines) shows that evictions are increasing dramatically. Between January and June 2004, 334,593 people were evicted in the urban areas of these countries. In January to June 2005, 2,084,388 people were evicted. The major reason for these evictions was the beautification of the city. In the majority of cases, people did not receive any compensation for the losses they incurred, and where resettlement did take place it was 25 to 60 kilometres from the city centre. (Ken Fernandes; *Some Trends in Evictions in Asia*, ACHR, March 2006).

the lower-middle-income groups have also been adversely affected and can no longer afford to purchase or rent a house in the formal land and housing market. They are also relocating to the peri-urban areas.

Due to relocation, transport costs and travel times to and from work have increased considerably. This has resulted in economic stress and social disintegration as family members who are earning have less time to interact with their families. Incomes have been adversely affected, since women can no longer find work in the relocation areas and children can no longer go to school.[3] Interviews with commuters in Bangkok by the author and in Karachi by the URC indicate that people living in locations far from their places of work (as more and more do) can spend up to six hours of travelling per day.

Due to an absence of alternatives for housing, old informal settlements have densified and, as such, living conditions in them have deteriorated, in spite of the fact that many of them have acquired water supply and road paving and have better social indicators, such as higher literacy, and better infant mortality rates.[4]

Local governments in all the research cities have evolved an "image" for their cities. This image is all about catering to the automobile, high-rise construction and the gentrification of poor areas. For this they are seeking foreign investment for building automobile-related infrastructure and elite townships. Much of this is being implemented through the BOT process which is two to three times more expensive than the normal local process of implementation. Foreign investment has also introduced foreign fast-food outlets, stores for household provisions, mobile phone companies, expensive theme parks and golf courses. Much of this development has pushed small businesses out of elite and middle-income areas and has occupied the public parks and natural assets of these cities for elite and middle-middle-class entertainment and recreation at the expense of low- and lower-middle-income communities.

An increase in the number of automobiles in Asian cities has created severe traffic problems and this in turn increases time taken in travel and in stress- and environment-related diseases.[5] Much of the financing of automobiles

---

[3] In Karachi, due to the relocation of over 14,000 households for the building of the Lyari Expressway, the schooling of more than 26,000 children has been disrupted. (*Lyari Expressway: Citizens' Concerns and Community Opposition*; Urban Resource Centre, Karachi, 2005.) In the Philippines, they have decided that evictions will only take place after the final exams have taken place in schools.

[4] In Pune (India), in the settlements surveyed for the report, densities in the last 25 years have increased by over 300 per cent without any major improvement in infrastructure and housing, resulting in massive environmental degradation and deterioration in living conditions.

[5] For instance, 81 per cent of children under five in Karachi develop acute respiratory diseases, mainly because of transport-generated air pollution. Studies of policemen and school children working or studying in traffic congested areas have dangerously high blood lead levels (IUCN; *Sindh State of Environment and Development*, IUCN Pakistan, Karachi, 2004).

is being done by loans from banks and leasing companies.[6] New transport systems (such as light rail) that have been or are being implemented do not serve the vast majority of the commuting public and, in most cases, are far too expensive for the poor to afford.[7]

As a result of the culture of globalisation and structural adjustment conditionalities, there are proposals for the privatisation of public sector utilities and land assets. In some cities the process has already taken place. There are indications that this process is detrimental to the interests of the poor and disadvantaged groups and there is civil society pressure to prevent such privatisation and to reverse it where it has taken place.[8] An important issue that has surfaced is as to how the interests of the poor can be protected in the implementation of the privatisation process.

The culture of globalisation and structural adjustment has also meant the removal or curtailing of government subsidies for the social sectors. This has directly affected poor communities who have to pay more for education and health and this in turn adversely affects their mobility through public transport.[9] In addition, the private sector in education, both at school and university levels, has expanded creating two systems of education, one for the rich and the other for the poor. This is a major change from the pre-1990s era and is having serious political and social consequences, as it is further fragmenting society into rich and poor sections.

As a result of the changes described above, there has been an enormous increase in real estate development. This has led to the strengthening of the nexus between politicians-bureaucrats and developers due to which building bye laws and zoning regulations have become easier to violate and due to which the natural and cultural heritage assets of Asian cities are in danger or in the process of being wiped out.

[6] For example, 502 vehicles have been added to Karachi per day during the last financial year. It is estimated that about 50 per cent of these have been financed through loans from banks and leasing companies, who have never had as much liquidity as they have today. This means that loans worth US $1.8 billion were issued for this investment, which could easily has been utilised for improving public transport systems.

[7] Cities such as Bangkok, Manila and Calcutta have made major investments in light rail and metro systems. Other Asian cities are following their example. However, these systems are far too expensive to be developed on a large enough scale to make a difference. Manila's light rail caters to only 8 per cent of trips, Bangkok's skytrain and metro to only 3 per cent of trips and Calcutta's metro to even less. The light rail and metro fares are three to four times more expensive than bus fares. As a result, the vast majority of commuters travel by the run-down bus system (for details, see Geetam Tiwari; *Urban Transport for Growing Cities*, Macmillan India, New Delhi, 2002 and Arif Hasan; *Understanding Karachi's Traffic Problems*; Daily Dawn, January 29, 2004).

[8] The privatisation of Manila's Water Supply System has benefited the rich and upper-middle-income areas and has had an adverse effect on lower-middle-income and lower-income areas. The privatisation of Karachi Electric Supply Corporation has created immense problems of power distribution and there is now public pressure to de-privatise it.

[9] Aquila Ismail; *Transport: URC Karachi Series*, City Press, Karachi, 2002.

There are multiple agencies involved in the development, management and maintenance of Asian cities. In most cases, these agencies have no coordination between them. In addition, in most cities there are central government interests that often override local interests and considerations. This makes sustainable planning and its implementation, especially related to environmental and land-use issues, problematic.

In all cities but two, governments that are already heavily in debt are seeking loans from international financial institutions (IFIs). Development through these loans is exorbitantly expensive and loan conditionalities are detrimental to the development of in-country technical and entrepreneurial expertise and to the evolution of effective municipal institutions.[10]

In most cases IFI and bilateral agency-funded projects seldom have any coordination between them and/or with national programmes, resulting in duplication and a waste of resources.[11]

- In all the case study cities, there has been a process of decentralisation. This has opened up new opportunities for decision making at the local level and for the involvement of local communities and interest groups in the decision-making process. In some cases, this has also meant a weakening of the community process in the face of formal institutions at the local level. In many cases the buffer of the bureaucracy between the elected representatives and the people has been removed giving a *carte blanche* to the elected representatives to take decisions without consultation with anyone. In this regard the synthesis paper asks two important questions "Does decentralisation give city governments more power and resources and thus capacity to act?" and "If city government does get more capacity to act does this actually bring benefits to urban poor groups?"

---

[10]According to research carried out by the OPP in Karachi, the government develops infrastructure at four to six times the cost of labour and material involved. When loans are taken from IFIs, the cost goes up by 30 to 50 per cent due to foreign consultants and related purchase conditionalities. Where an international tender is also a conditionality the cost can go up by an additional 200 to 300 per cent. Thus something whose cost is US $1 in material and labour terms is delivered at a cost of US $20 to US $30. According to a paper by the Cambodia Development Resource Institute titled "Technical Assistance and Capacity Development in an Aid-Dependent Economy, Working Paper 15, Year 2000", in 1992, 19 per cent of all aid money was spent on technical assistance. In 1998, it had increased to 57 per cent.

[11]For example, there are currently seven IFI-funded studies being carried out for Karachi, which have little or no interaction with each other. These studies will result in a proposal for IFI loans. According to a report "The Asian Development Bank: In Its Own Words: An Analysis of Project Audit Reports for Indonesia, Pakistan and Sri Lanka, prepared by ADB Watch, July 2003", over 70 per cent of all projects funded by the ADB have been unsuccessful and/or unsustainable.

THE CURRENT KARACHI CONTEXT

In recent decades, the whole approach to planning has undergone a change in Karachi. The local government is obsessed with making Karachi "beautiful" to visitors and investors. As a result, it has adopted the following thinking, which the author has experienced in other cities such as Manila, Almaty, Phnom Penh, Ho Chi Minh City, as well as in conversation with delegations from India.

Karachi has to be a "world-class city". What this actually means has never been explained, but it is one of the objectives of the Karachi Master Plan 2020.

The city has to have "investment-friendly infrastructure". Again, what this means has not been clearly defined. However, it seems from the programmes of the local government that this means the following:

- Flyovers and elevated expressways as opposed to traffic management and planning
- High-rise apartments (eight floors is considered to give a proper image for the city) as opposed to upgraded settlements
- Malls as opposed to traditional markets (which are being removed)
- Removing poverty from the centre of the city to the periphery to improve the image of the city, so as to promote direct foreign investment
- Catering to tourism rather than supporting local commerce
- Seeking the support of the international corporate sector (developers, banks, suppliers of technologies and the IFIs) for the above

The above agenda is an expensive one. For this, sizeable loans have been negotiated with the IFIs on a scale unthinkable before.[12] Projects designed and funded through previous loans for Karachi have all been failures.[13] Given this fact and the fact that local government institutions are much weaker in technical terms than they were in previous decades, it is unlikely that the new projects will be successful. Also, it is quite clear from the nature of projects being funded that they are not a part of a larger planning exercise. In addition, there are also projects that are being floated on a BOT process. They are also not a part of any plan. It is obvious that projects have replaced planning and that the shape of the city is being determined increasingly by local and foreign capital and its promoters and supporters. This agenda is also anti-people and has resulted in increased numbers of

[12]Between 1976 and 1993, the Sindh province in which Karachi is located borrowed US $799.64 million for urban development. Almost all of this was for Karachi. Recently, the government has arranged to borrow US $800 million for the Karachi Mega City Project. Of this, US $5.33 million is being spent on technical assistance being provided by foreign consultants.

[13]ADB-793 PAK; *Evaluation of KUDP and Peshawar Projects*, 1996.

evictions both of settlements and of hawkers, and the creation of conditions that make it difficult for working-class people to access previously accessible public space. As a result, multi-class public space for entertainment and recreation is rapidly disappearing in Karachi.

CIVIL SOCIETY RESPONSE IN KARACHI TO THE NEW DEVELOPMENT PARADIGM

Two civil society organisations in Karachi have been involved, along with a network of community organisations and NGOs, in dealing with the issues described above that adversely affect poor communities and society as a whole. One of these organisations is the Orangi Pilot Project (OPP) and the other is the URC, both in Karachi. This chapter will not describe the work of the OPP in detail since it has been presented at many international forums and is well known. However, a brief synopsis is given below. The OPP:

- promotes the upgrading of informal settlements through community mobilisation, finance and management by providing technical advice and managerial guidance to communities
- establishes partnerships with government agencies whereby they develop the off-site infrastructure and the communities finance, develop, manage and maintain the on-site infrastructure
- runs an education programme which encourages educated young women and men to open informal schools in low-income settlements which become formal schools through a process of teachers' training and upgrading
- operates a savings and credit programme for establishing rural and urban cooperatives
- reaches out to over two million population at 248 locations in Pakistan in 11 Pakistan cities. This network supports the advocacy work of the URC

The objectives of the OPP were not to propose alternatives to overall urban development planning but, due to the adverse effects of the new paradigm on poor communities, OPP's recent work has focused on trying to mitigate the effects of this paradigm and to support the URC's advocacy work.

THE URBAN RESOURCE CENTRE

The Urban Resource Centre (URC) was set up in 1989 by teachers of architecture and planning, NGO activists and community leaders. It has a five full-time staff members supported by six to seven interns at any given time.

The community organisations and networks developed by the OPP have become an integral part of it. The basic objective of the URC is to influence the planning and implementation process in Karachi to make it more environment- and poor-friendly by involving communities and interest groups in this process. To further its objective the URC carries out the following activities:

- It collects information regarding the city and its plans and disseminates it to the media, NGOs, CBOs, concerned citizens and formal and informal interest groups. To this end it keeps files of news clippings on all major Karachi issues and these are available to researchers, students and the media.
- It analyses local and federal government plans for the city from the point of view of communities (especially poor ones), interest groups, academia and NGOs. These analyses are done with the involvement of interest groups and low- and lower-middle-income communities and are carried out through a process of public forums, to which government planners and representatives of development agencies are invited along with the media. The forums are documented, published in the media and become a basis for public debate and discussion. The more important issues are developed into promotional and advocacy literature. On the basis of forums, the URC attempts to arrive a larger consensus.
- It identifies and promotes research and documentation on major issues in Karachi and it monitors developments and processes related to them.
- It seeks to create professionals and activists in the NGO/CBO and government sector who understand planning issues from the point of view of local communities, especially poor ones. To make this possible it operates a Youth Training Programme, whereby it gives one-year fellowships to young university graduates and community activists who help it with research, documentation and interaction with communities and interest groups. Through these fellowships the URC seeks to broaden its base in society as a whole.
- It arranges lectures by eminent professionals and experts on national and international development-related issues, which are attended by grass-roots activists, NGOs, government officials, academia and representatives of interest groups. This helps organisations and individuals to relate their work to larger national and international issues.
- It promotes and supports a network of CBOs and NGOs for networking on major Karachi-related development issues and projects.
- It monitors and documents evictions, identifies vulnerable communities and informs them of possible threats to them, and publishes on eviction issues which in turn get taken up by the print and electronic media.

Much of the URC's work has been related to the transport sector, since the absence of facilities and the new proposals being floated affect both the environment and the lives of the majority of Karachi's population, 72 per cent of whom belong to the lower- and lower-middle-income class. Through news items the URC was able to identify and initiate a dialogue with the major informal actors in the Karachi transport drama. These included the Karachi Bus Owner's Association, Transport Ittehad, the Mini Bus Driver's Association and a number of consumer organisations. The URC's forums linked these organisations with government planning agencies. This association has grown. In addition, the URC has held discussions with the wholesale market operators and small manufacturers in the inner city of Karachi, which is congested and environmentally degraded. Based on these discussions and the consensus arrived as a result, and on documentation of the inner city developed by the Department of Architecture & Planning at the Dawood College, Karachi, it has advocated the relocation of these markets to the Northern Bypass. The city government is now making proposals for relocation of these markets.

THE KARACHI DEVELOPMENT AUTHORITY'S (KDA) MASS TRANSIT PROPOSAL AND THE DEVELOPMENT OF THE URC MODEL OF RESEARCH AND ADVOCACY

The first major initiative of the URC became a model for its research and advocacy work. This initiative was a questioning of the Karachi Mass Transit Project, which was developed in 1994, with advice from World Bank consultants, by the Karachi Development Authority's Mass Transit Cell. The project proposed six corridors of elevated light rail transitways. Corridor One (15.4 kilometres) was to be implemented in Phase 1 at a cost of US $668 million. After studying the proposal the URC raised the following objections.

- The elevated transitways would be an environmental disaster in the inner city and would also adversely affect the built heritage of the inner city.
- Corridor One generated no commuters itself. Commuters came to Corridor One from other locations.
- An abandoned circular rail corridor already existed. Its revival and extension would cost a fraction of the cost of Corridor One and would serve a far larger area than the six corridors put together.

The city planners refused to accept the URC's objections as valid. The URC then initiated the formation of the Citizens' Forum on Mass Transit

(CFMT), which consisted of professionals, NGOs, concerned citizens, media organisations and the OPP community networks. The URC also held meetings with communities along the corridor to explain the project to them. They were horrified and expressed their concerns to their elected representatives and through letters in the press. After considerable weight of opinion had been built in the media for and against the project, the CFMT held a citizens' forum and presented its findings to them along with alternatives to the government plan. The forum was reported extensively in the media and a major debate in and outside the media ensued. This resulted in major changes in the Karachi Mass Transit Project and the development of proposals for the revitalisation and extension of the abandoned circular railway.

The most important part of this process was the creation of a large informal network consisting of NGOs, community organisations, media, concerned citizens, professionals, academic institutions and central government departments. This network becomes operative whenever the need arises. (See Appendix 3 for institutions linked to the network.) It has also resulted, after considerable disagreement, in a very cordial and supportive relationship between the URC and the Karachi Mass Transit Cell of the city government.

The above procedure was adopted successfully by the URC to question a number of government initiatives. These include:

The URC challenged the US $100 million Asian Development Bank (ADB)-funded Korangi Waste Water Management Project for Karachi and, along with the OPP and other NGOs and community organisations, pushed for the US $20 million alternative prepared by the OPP. As a result, the ADB loan was cancelled in favour of the OPP alternative. This process also led to the creation of the Water and Sanitation Network in Karachi, which monitors water and sanitation projects proposed by government agencies.

The privatisation of the Karachi Water and Sewerage Board (a government organisation) was opposed by civil society organisations along with the URC. This opposition led to the creation of a government-appointed review committee on the basis of whose report the privatisation process was cancelled.

The URC's monitoring of forced evictions and the holding of forums around them has created an awareness in society regarding their negative aspects. As a result, political parties have come together recently to oppose evictions proposed by the local government and this has led to their cancellation.

The URC's research and forums on solid waste management (which was carried out in collaboration with scavengers and informal solid waste recyclers) has led to the informal recycling industry being accepted as an important interest group in this sector.

THE NORTHERN BYPASS AND
THE LYARI EXPRESSWAY

The above model of negotiation and consensus-seeking was not followed by the URC in the case of the Lyari Expressway. The project and the manner in which changes were sought in it are explained below.

The northern and southern bypasses were proposed by the Karachi Development Plan 1975–1985. They were to be built from the port to the Super Highway and the National Highway respectively, and, as a result, all port-related traffic would be able to bypass the city. The southern bypass could not be built because of opposition from the Defence Housing Authority (DHA) through which a part of its proposed route passed. The DHA was concerned about the environmental pollution that the bypass would cause to some of its neighbourhoods. It was also politically powerful enough to get its point of view accepted. The northern bypass was not built either for a variety of political and resource reasons.

In 1986, a group of public-spirited citizens proposed the Lyari Expressway as an alternative to the northern bypass. This Lyari Expressway proposal consisted of building a road from the port along the Lyari River to the Super Highway, which is Karachi's main link with the rest of Pakistan. A government study found the construction of the Lyari Expressway unfeasible if routed along the river banks as over 100,000 people (at that time) living along the river would have to be evicted as a result of its construction. However, the idea of the Expressway appealed to the politicians and planners and so in 1989 the KDA involved the Canadian International Development Authority (CIDA) in the Lyari Expressway project. CIDA proposed an elevated corridor (on columns) in the middle of the river as the most feasible option as it would not displace any Lyari Corridor communities. However, in 1993, rains flooded the lower-lying Lyari Corridor settlements. As a result, planners proposed the building of the Lyari Expressway along both banks as a solution for flood protection and also for generating funds through a toll for cost recovery. The skyway project, however, remained unaffected.

The URC objected to both the proposals, for reasons that are explained later in the text, and expressed its point of view through a number of forums and newspaper articles. However, there was little or no response from politicians and government planners on the concerns raised by the URC.

The URC then held meetings along the Lyari Corridor and explained the Lyari Expressway project to the communities. As a result, the Lyari Nadi Welfare Association (LNWA), consisting of 42 community organisations, was formed. Meanwhile, the URC also developed alternative plans for redirecting port traffic from the port to the Super Highway and costed the plans. These plans, along with photographs, maps and estimates, were given to the LNWA and they, in turn, contacted their MNAs and MPAs and the chief minister of Sindh. As a result, the project was delayed. All this information and documentation was also sent to CIDA and the Canadian Embassy. CIDA finally backed out and the skyway proposal was shelved.

In 1994, the Karachi Metropolitan Corporation (KMC) decided to build the Expressway on either side of the river. Eight thousand shacks and small business enterprises at the lower end of the river were removed for its construction. However, the demolition work along the Lyari River led to opposition to the project by citizens, NGOs and the more consolidated, and comparatively politically powerful, Lyari communities, due to which a number of politicians became concerned. This opposition led to public hearings in 1996, which were arranged by the senior minister of the Sindh government. As a result of the public hearings, it was decided to build the northern bypass and abandon the building of the Lyari Expressway. Work on the northern bypass was begun. However, in June 2001, the government decided to build both the northern bypass *and* the Lyari Expressway, in violation of the decisions taken as a result of the 1996 public hearings.

The government has justified the project by saying that it will ease traffic flow within the city and will also remove people from the flood zone to safer locations.

Citizens and community concerns

A number of NGOs and academics related to urban planning got together with the Lyari community organisations and drew up a list of concerns. These were sent to the president, the city Nazim and all other relevant agencies and government departments through the URC. No reply to these concerns was given by any individual or agency. The six major concerns are summarised below.

The Expressway project is not a part of a larger city planning exercise. There are cheaper and easier methods for easing traffic flow in Karachi that have been proposed repeatedly by the KDA's Traffic Engineering Bureau, Karachi academics and professionals. In addition, the building of the Expressway does not solve the major environmental problems of the city or of the areas it passes through. These problems can only be solved by the relocation of the major wholesale markets and congested informal industrial activity and warehousing in the inner city.

If the Expressway is going to be used for heavy port-related traffic, it will cause severe environmental pollution and hence further degradation along the already densely polluted Lyari Corridor. On the other hand, if the Expressway is to be used only for intracity traffic, a different sort of land-use change will occur. In this case, there will be a sharp increase in land values, which will lead to the eviction of the remaining old settlements along the Corridor. In either case the development along the corridor will eventually add approximately 100,000 vehicles per day on a corridor that passes through the centre of the city and is vehicle free today.

The building of the Expressway will lead to the demolition of a lot of the city's cultural heritage, 25,400 houses and about 8,000 commercial and manufacturing units. The education of 26,000 students will be affected and more than 40,000 wage-earners will lose their jobs. Five billion rupees of investments made by poor people in homes and in acquiring legal infrastructure connections will be destroyed, and centuries-old communities will be dislocated.

The acquisition of land for the Expressway was violating Pakistan's environmental laws, the provisions of the Land Acquisition Act and the Global Plan of Action of UN Habitat–II 1996 and other international covenants to which Pakistan is a signatory.

The proposed resettlement plan was flawed since it left the affected families poorer than before, unemployed, without utilities and without health and education facilities, all of which they enjoyed before.

The project was designed with consultation with the communities, or with the academic, professional and NGO lobbies that opposed it, although Karachi professionals proposed alternatives to the Expressway design that addressed some of the concerns raised by the URC.

Community opposition to the Expressway

Community opposition to the Expressway has been well organised. It has been supported by the Action Committee for Civic Problems (ACCP), a Karachi-based NGO. The ACHR contacted its international network who wrote over 1,200 protest letters to the Karachi Nazim. This resulted in a UN-backed ACHR Fact-Finding Mission to Karachi which met elected representatives and bureaucrats at the federal and provincial level. The government-approved rehabilitation plan for those affected is the result of this Mission and the organised opposition of communities to this project.

Lessons learned

Two important lessons have been learnt from the Lyari Expressway experience.

The legal profession, bureaucrats and planners are not aware of the international covenant that the government has signed. Implementation of these covenants would lead to the development of a far more humane social and physical environment.

Politicians and government officials are very sympathetic to the needs of the environment and to the exclusion of low-income communities from the city. However, planners have been trained to deliver a form of development that is anti-pedestrian, anti-street, anti-dissolved space and anti-mixed land use. The fault lies in their basic education, which does not give them a structure of thinking that makes innovation possible, and also in the fact that the existing building bylaws and zoning regulations support this form of planning.

URC'S NEW INITIATIVES

The URC is currently involved in a number of new initiatives. A brief description of the important ones is given below.

- Local government and the Defence Housing Authority (an elite housing colony) have banned hawkers, push-carts and jugglers from the beaches in their jurisdiction. The URC has taken up this issue through the media and forums. As a result, the resident's association of the DHA has intervened to permit the hawkers and others to ply their trades on the beach. It is important that an elite organisation has intervened to protect and promote the interests of the poor. The city government has yet to accommodate hawkers on its part of the beach.
- The government agencies have decided to privatise 14 kilometres of beach in Karachi, which is currently used for recreational purposes by the citizens (especially the poor) of the city. The privatised zone will have high-income condominiums, five-star hotels and up-market eating places and clubs. This development is being carried out in collaboration with a Dubai-based investment company and a Malaysia-based construction company. The URC is currently mobilising public opinion against this proposal which it considers to be a usurpation of public space, and hopes to hold an informed forum on it. So far over 400 NGOs, CBOs and schools from all over Karachi and Pakistan along with over 5,000 individuals, have signed a petition opposing this US $1,500 million project. The project will also bring an additional 25,000 car trips per day to the beach. The planners plan to solve this problem by building underpasses, flyovers and multi-storey car parking.
- Hawkers exist all over Karachi. More than 3,500 of them work in the city centre. The government keeps evicting them and they keep coming back. The URC, in collaboration with hawkers and residents of the area, has worked out a plan for the rehabilitation of the city centre hawkers on pedestrianised streets, near bus stops and transport terminals. The plans were discussed with the previous Nazim and will be discussed with the present city government officials. The reasons for the failure of hawkers' rehabilitation projects in the past has been that the relationship between the hawkers and the commuter public, bus stops and public transport terminals has not been properly understood and accommodated in the plans. Additionally the relationship between the formal sector retail outlets, the hawking business and governance issues has not been understood either. The URC's research work has identified and documented all these relationships.

The URC has been monitoring evictions in the city since 1992. It has succeeded in making this an issue with the media, political parties and other civil society organisations and also with certain sections of the elite. It has provided settlements along the railway tracks, where the occupants are likely to be evicted, with videos and details showing that formal sector buildings (such as factories, middle-income flats and housing and government institutions) are also encroaching on the railway's "right of way". The organisations have lobbied with this information and demolitions have been curtailed as a result. It is hoped that with this information the "right of way" will be reduced preventing the eviction of tens of thousands of households.

The URC has also initiated a Secure Housing Initiative (SHI) whereby settlements under threat are documenting their history, government and community investments in their infrastructure, issues related to land title, and details regarding the families living in them. This information will be used for lobbying against evictions and for developing support for the SHI.

A number of professionals and activists in other cities in Pakistan wish to establish URCs. Centres in Lahore, Faisalabad and Rawalpindi are already in existence. The Karachi URC has been supporting these initiatives through orientation and training in which the organisations visit the Karachi URC and the Karachi URC visits them. A number of joint initiatives are in the offing.

The URC continues to monitor government plans related to solid waste management, the Lyari Expressway, transport and mass transit issues and the Karachi Master Plan. It continues to operate its forums and networking, institutional development of CBOs, carrying out research, producing publications and developing its library for use by various interest groups.

CONCLUSIONS

The URC process has created an awareness in the public, the media and the government regarding the problems official planning can create for communities and weaker interest groups. It has also created an understanding of the negative aspects of IFI-funded projects and the possible alternatives to them that rely entirely on local resources. The process has also created a link between society, media and the government on urban planning issues. The media and the government (even if it does not agree with the URC analyses) consults with the URC. This has generated discussion and debate on issues that were never discussed before in the media, in the government planning institutions and in the institutions where government bureaucrats are trained. Most of the URC's work is on transport- and eviction-related issues, the two of which are closely interrelated. This interrelation is seldom understood or catered to in official planning.

The URC is essentially trying to create a space of interaction between politicians, planners (government agencies, academic institutions) and people (communities and small formal and informal interest groups). This space has been created but it needs to be nurtured and subsequently institutionalised. The diagram in Appendix 4 illustrates this concept.

The URC has been constantly accused by local government, consultants and contractors (both national and international) as being anti-development. Also the URC understands that it cannot fight international capital and the forces that are determining the shape and form of our cities. Therefore, it has developed a four-point city planning agenda which it hopes to promote. This is given below.

- Planning should respect the ecology and the natural environment of the region in which Karachi is located.
- Land use should be determined on the basis of social and environmental considerations and not on the basis of land value or potential land value alone.
- Planning should give priority to the needs of the majority of the population who, in the case of Karachi, belong to the lower-income and lower-middle-income classes, the majority of whom are pedestrians, commuters, informal settlement dwellers and workers in the informal sector.
- Planning should respect the tangible and intangible cultural heritage of Karachi and of the communities living in it.

The implementation of the above agenda can only take place if a large network of civil society organisations, academics, professionals and community organisations support it. If implemented, it will bring about major changes in transport-related planning.

APPENDIX 1: NAMES OF RESEARCHERS AND TITLES OF THE REPORT

- **Beijing:** Alexander Andre, Yutaka Hirako, Lundrup Dorje and Pimpim de Azevado (2004), *Beijing Historic Case Study*;
- **Pune:** Bapat Meera (2004), *Understanding Asian Cities: The Case of Pune*;
- **Chiang Mai:** Charoenmuang, Duongchan, Apavatjurt Tanet Charoenmuang, Wilairat Siampakdee, Siriporn Wangwanapat and Nattawoot Pimsawan (2004), *Understanding Asian Cities: The Case of Chiang Mai*;
- **Phnom Penh:** Crosbie, David (2004), *Understanding Asian Cities: Phnom Penh, Cambodia*;

- **Karachi:** Hasan Arif and Asiya Sadiq (2004), *Understanding Asian Cities: The Case of Karachi*;
- **Muntinlupa:** Karaos, Anna Marie and Charito Tordecilla (2004), *Understanding Asian Cities: The Case of Manila, Philippines*;
- **Hanoi:** Thi Thu Huong, Nguyen (2004), *Understanding Asian Cities: The Case of Quynh Mai Ward, Hai Ba Trung District, Hanoi, Vietnam*;
- **Surabaya:** Johan Silas, Andon, Hasian and Wahyu, the Laboratory for Housing and Human Settlements, ITS, Surabaya (2004), *Surabaya and People's Role*.

APPENDIX 2: UNDERSTANDING ASIAN CITIES RESEARCH: POSSIBLE DIRECTIONS (JUNE 30, 2003)

A. OBJECTIVES

To understand the process of socio-economic, physical and institutional change in Asian cities, the actors involved in it and its effect on disadvantaged communities and interest groups.

To identify/understand civil society and/or community movements and their role in the process of change.

To help the NGO, CBO, ACHR partners/ACHR in taking a position on national and international forums on housing rights and development issues.

To support in eight cities a group that monitors the city/continuous learning.

B. RESEARCH OUTLINE

1. Demography/census data
    a) Trends established by comparing two or three census:
        - Population (male/female)
          Age and sex pyramid analysis
        - Married/divorced population
          different age groups (male/female)
        - Literacy/educational attainment
          different age groups (male/female)
        - Source of information (if TV, is satellite/cable available)
        - Telephones per 1,000 population
        - Employment
          different age groups (male/female)
          formal-informal/categories
    b) Trends established by comparing two or three housing census or any other secondary data over different periods:
        - Nature of housing stock/congestion indicators
          (persons per room/rooms per person)
        - Availability of physical infrastructure
          (electricity, water, sewage, drainage, road paving, gas)
        - Tenure security? Squatter households (total and percentage)

c) Select two settlements: one consolidated and one comparatively new and through a survey of 100 houses in each settlement develop data for:
   - Incomes
   - Employment
   - Mode of transport and cost to and from work
   - Literacy
   - Source of information
   - Security of tenure
   - Nature of housing stock and how the houses were built
   - Availability of infrastructure
   - Effect on respondents due to environmental degradation at city level
     (noise, air, solid waste, water, sewage disposal, bad transport)
d) Identify and establish contacts with organisations dealing with the above issues and acquire their literature.
e) Conclusions (conclusions for all sections can be established through in-house discussions/workshops).

2. **Poverty profiles and funds**
   a) Profiles:
      - Most countries/cities now have poverty profiles. These could be reproduced/definition of poverty.
      - Causes and repercussions of the indicators of poverty (you could conduct a workshop or discussion to reach conclusions).
   b) Poverty alleviation funds (most countries have them now) or other similar funds:
      - Scale of the funds
      - Source of the funds
      - Institutional arrangement for the funds/functions
      - Utilisation so far/achievements
      - A critical analysis.
   c) The debt situation:
      - The nature, scale and repayment of debt over three different periods in the last 20 years
      - What has it been utilised for so far?
   d) Conclusions.
   e) Identify relevant persons/organisations and acquire all relevant literature.

3. **The institutional set-up**
   a) Descriptive: from secondary sources, personal knowledge:
      - The structure of local government and planning institutions and their relationship with each other. A critical analysis.

- How plans are made, decisions taken regarding them, implemented and financed. Examples would be welcome.

b) Case study comparison of process, costs and effects (social and political) between an:
- IFI-funded infrastructure project
- A local government-funded project
- A community project and the reasons for the differences.

c) Decentralisation:
- The new decentralised system: a description
- Weaknesses, strengths and potential.

d) Conclusions

e) Identify persons/organisations and acquire all relevant literature.

4. **The physical city (secondary data, observations, in-depth interviews)**

a) Physical growth over time (maps of three different decades):
- Establish trends related to land use, housing, informal settlements, formal and informal industry and commercial activity.
- Some understanding of environmental issues and their causes related to noise and air pollution, sewage and waste water disposal and solid waste management systems.
- Location of environmentally degrading activity and the reasons for those locations
- Cars, motorcycles, public transport seats (total and per 1,000 population)
- The effect of the above on poor communities.

b) Does your city have a development/master plan? If yes,
- What are the salient features?
- How is it different from previous plans?
- Who is funding it and/or providing technical support?
- Which local interest groups/academic institutions are involved in it?
- Are there any "poor poor" elements in it?
- What are the problems in implementing the plan? Identification and discussion.

c) Who owns land in your city and scale of ownership
- Government institutions
- Private sector
- Individually
- Others

d) Three important development projects over the last two decades
- Description and justification
- Cost and source of funding

- Was there any opposition to the projects? If so, why and by whom, and what happened?
- Benefits and/or disadvantages for the poorer populations
- Were there alternatives? If so, what were those advantages?

e) Three important projects being currently planned or executed:
- Description and justification
- Cost and source of funding
- Was there any opposition to the projects? If so, why and by whom, and what happened?
- Benefits and/or disadvantages for the poorer populations
- Were there alternatives? If so, what were those advantages?

f) Identify (or in your opinion) three major physical initiatives that would benefit the urban poor in your city. Can you give an approximate cost?

g) Establish links with relevant organisations and acquire all relevant documents, plans and reports.

h) Conclusions.

5. **Housing policies and programmes**
a) The demand-supply gap:
- In three different periods over the last 25 years.
- How is it taken care of?

b) Informal/squatter settlements:
- Scale in three different periods over the last 25 years
- Indication of change of locations of development
- The relocation issue: scale in three different periods over the last 25 years in real and percentage terms
- Evictions: scale over different periods and reasons for the evictions
- Laws and procedures related to evictions and relocation (de facto and de jure).

c) Government housing policies:
- Current policies: What are they?
- How do they differ from previous policies?
- Their relevance to poor communities
- A critical analysis from the point of view of disadvantaged groups.

d) The issue of built-heritage.

e) Conclusions.

f) Identify relevant agencies and individuals and acquire all relevant literature on the subject.

6. **Civil society organisations**
- A critical analysis of different types (including interest groups based on trades such as hawkers, shopkeepers, transporters' organisations)
- Strengths, weaknesses, constraints

        – Nature of support/networking required to strengthen them
        – Conclusions
        – Identify persons, acquire all relevant literature.

7. **The impact of globalisation (and its culture) and structural adjustment**
   a) Observations (also secondary sources):
      – On the younger generation
      – The changes in the cityscape, land-use patterns, entertainment and social relationships
      – The role of IT, satellite TV, pop culture
      – The effect of a gap between aspirations and means (crime?)
      – The extended family under stress and its repercussions.
   b) The economic impact (secondary sources):
      – Privatisation of water/sewage, solid waste, education, health
      – Cutbacks in public spending on social sectors
      – Effect on employment and its causes and repercussions
      – The increase in cost of urban utilities and its repercussions.
   c) Case studies to illustrate the above? Suggestions welcome.
   d) Conclusions.
   e) Acquire all relevant literature, identify organisations and individuals.

8. **Conclusions**

APPENDIX 3: THE EMERGING KARACHI NETWORK

A. **NGOs**
   1. Orangi Pilot Project-Research and Training Institute
   2. Orangi Charitable Trust
   3. Aurat Foundation
   4. Shirkatgah
   5. Citizen's Committee for Civic Problems
   6. Human Rights Commission of Pakistan
   7. Urban Working Group
   8. Pakistan Institute of Labour Education and Research
   9. Shehri
   10. Saiban
   11. Urban Resource Centre

B. **38 CBOs**

C. **Media organisations**
   1. Jung Forum
   2. ICN
   3. Press Club

    4. Manduck Productions

D. **Interest groups**
1. Minibus drivers associations
2. Transport Ittehad
3. Tanker Owners Association
4. Karachi Bus Owners Association
5. Solid waste recyclers associations (6)
6. Hawkers associations (8)
7. Kabari Welfare Anjuman
8. Scavengers associations

E. **Government departments**
1. Sindh Katchi Abadi Authority
2. City Government Mass Transit Cell
3. Karachi Public Transport Society
4. Sindh Cultural Heritage Committee
5. Karachi Master Plan Department

F. **Academic institutions**
1. Dawood College, Department of Architecture and Planning
2. NED University, Department of Architecture and Planning
3. Karachi University:
   – Department of Architecture and Planning
   – Social Works Department
   – Mass Communications

G. **National Institute of Public Administration**

Chapter 4

City design and transport
Observations at different urban scales

Philipp Rode and Ricky Burdett

CONTENTS

DENSITY CONFIGURATIONS

Figures 4.1 through 4.12 give residential density levels and GDP per capita, proportion of one-person households, car ownership per 1,000 persons, foreign-born residents and unemployment figures for six cities (Mexico City, Shanghai, Johannesburg, New York, London and Berlin) around the world. In this chapter, we discuss how city design affects transport patterns in urban areas.

THE LONDON CASE: ACCESSIBILITY AND DENSITY

The information below was generated by London School of Economics (LSE) research on 'Density and Urban Neighbourhoods in London'. It is based on a spatial and accessibility analysis for five higher-density neighbourhoods combined with socio-economic data derived from the UK Census, over 70 stakeholder interviews as well as a quantitative survey that included more than 1,900 local residents. For the purpose of this work, the specific relationship of accessibility and city design was re-examined. Accessibility is largely determined by two land-use characteristics: density and mixed use. The perceivable relationship of density and accessibility is best looked at by separating opportunities and constraints resulting from higher-density levels.

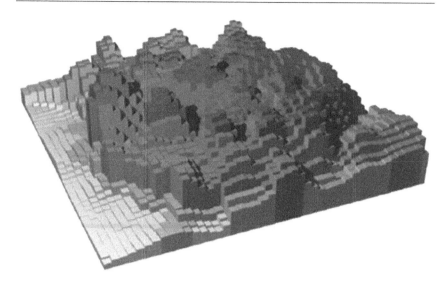

Figure 4.1 Mexico City, residential density levels.

Figure 4.2 Mexico City, key trends.

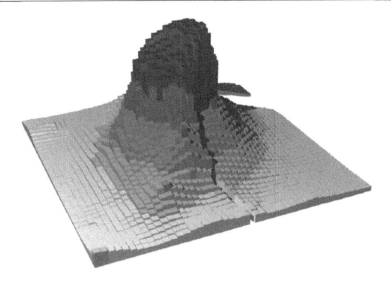

Figure 4.3 Shanghai, residential density levels.

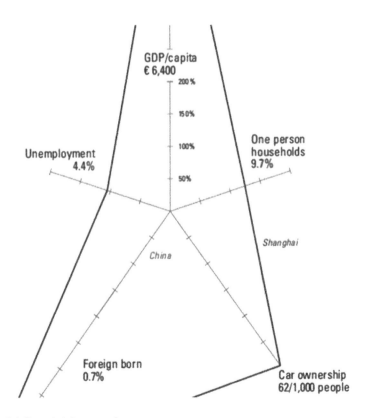

Figure 4.4 Shanghai, key trends.

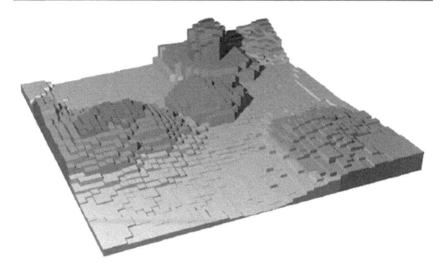

Figure 4.5 Johannesburg, residential density levels.

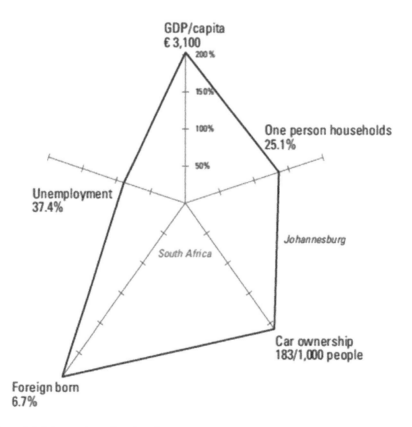

Figure 4.6 Johannesburg, key trends.

Figure 4.7 New York, residential density levels.

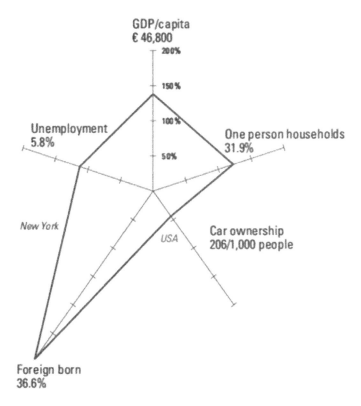

Figure 4.8 New York, key trends.

Figure 4.9 London, residential density levels.

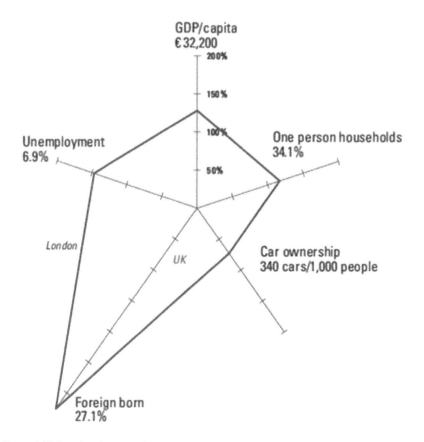

Figure 4.10 London, key trends.

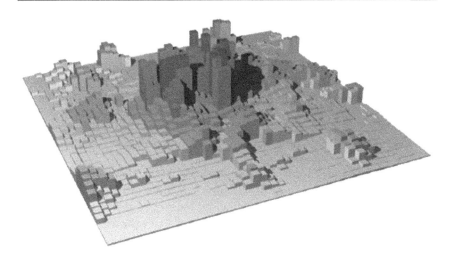

Figure 4.11 Berlin, residential density levels.

Figure 4.12 Berlin, key trends.

Opportunities for accessibility resulting from higher density

Higher densities come along with several opportunities regarding accessibility. The basic logic here is that a critical mass of residents leads to the physical proximity of origins and destinations. In addition, temporal proximity, as a result of a higher density of transport infrastructure (roads, walkways, bike paths, rail and waterways), allows for faster access to more distant destinations, which, in the case of public transport, also includes a higher service density and frequency.

This logic becomes evident in Figure 4.13, where the Public Transport Accessibility Level (PTAL) of all Greater London wards is represented in a scatter plot together with the residential density of these wards. PTALs measure the accessibility of the geographic centre of each ward to the public transport network, taking into account walk access time and service availability. A strong positive correlation emerges even when including areas of very good public transport in Central London with a high workplace density but low residential densities.

Figure 4.13 ties PTAL scores closely to gross residential density with a correlation factor of 0.520. Such a high correlation emphasises the degree of necessary interdependence between high residential density and public transport provision. Further, public transport provision, expressed in PTALs, illustrates how the supply side influences passengers' mode choices. On the basis of Greater London wards, PTALs and public transport trips to work correlate to a factor of 0.374.

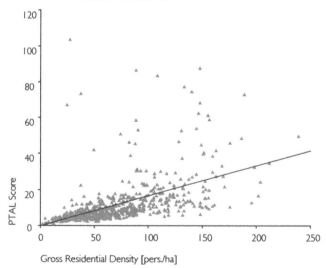

Figure 4.13 Scatter plot of PTALs and gross residential density. Significant correlation Pearson's coefficient: 0.520.

Source: Census 2001.

In addition, Figure 4.14 shows the centrality of public transport provision with increasing levels from the city fringe to the city centre, showing the best accessibility levels for the commercial and business centre and then for high-density residential areas. A higher density of public transport further results in a temporal proximity to a larger area as illustrated by Figures 4.15 and 4.16. These maps show areas that can be reached within a certain time from the Glyndon ward (Woolwich, Greenwich) characterised by a lower-density area along the Thames Gateway and Southfield ward (Ealing), being part of a higher-density corridor stretching west, respectively. Both are at a similar physical distance from Central London. It is clear that a much larger area is accessible to Southfield (high PTAL) within a given time than is the case for Glyndon (low PTAL). Consequently, for the five higher-density neighbourhoods analysed by the LSE, the survey was able to identify accessibility by public transport (43%) as the second

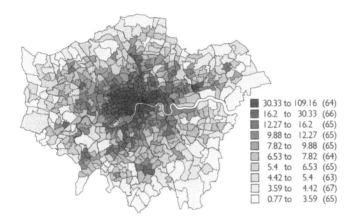

Figure 4.14 PTAL score.

Source: Transport for London 2003.

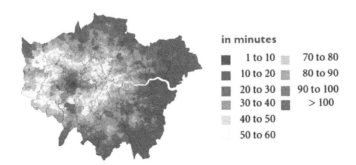

Figure 4.15 Temporal proximity Southfield.

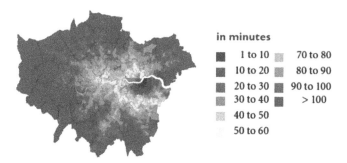

in minutes

■ 1 to 10   70 to 80
■ 10 to 20   80 to 90
■ 20 to 30 ■ 90 to 100
■ 30 to 40 ■   > 100
  40 to 50
  50 to 60

Figure 4.16 Temporal proximity Glyndon.

and quality of public transport (23%) as the fifth single best thing about these areas.

That higher density levels not only lead to a better provision of public transport, particularly the bus service, but are further responsible for certain mobility patterns is shown in Figures 4.17 through 4.19. Comparing the two wards Clissold, with a gross residential density of 148 pers./ha, and Bensham Manor, with 111 pers./ha, allows for the following observation. Bus service in Clissold is accessible at a short walking distance (to simplify only measuring the distance to the bus line and not considering the actual location of the bus stop) from 83% of the entire ward area, whereas in Bensham Manor from only 52%. The modal share of buses in Clissold follows the logic of this advantage being 30% compared to 17% in Bensham Manor. However, the availability of the train service in Bensham Manor triggers a high percentage of commutes by train of 20%, whereas in Clissold, lacking both train and tube services, these add up to 20%.

The provision of the train service is, to a lesser extent, a direct product of high density at the neighbourhood level, and relies more on a strategic location along a railway corridor.

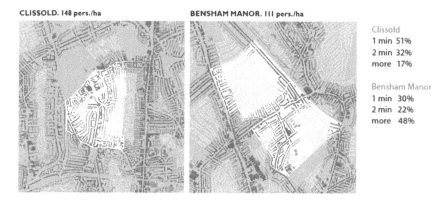

CLISSOLD. 148 pers./ha   BENSHAM MANOR. 111 pers./ha

Clissold
1 min 51%
2 min 32%
more   17%

Bensham Manor
1 min 30%
2 min 22%
more   48%

Figure 4.17 Bus access: 1- and 2-minute walk band.

The dependency of good public transport provision on high density in comparison to other modes of transport becomes even clearer when comparing different transport networks with each other. For the ward of Town, Figure 4.20 illustrates the transport network for pedestrian, car, bus and rail routes for an area of 2.25 km×2.25 km. For this relatively dense ward with 153 pers./ha, pedestrian and car routes offer fine-grained access

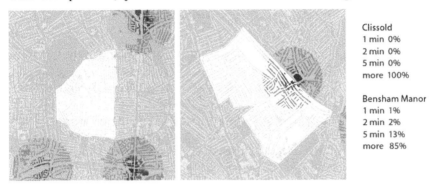

Clissold
1 min 0%
2 min 0%
5 min 0%
more 100%

Bensham Manor
1 min 1%
2 min 2%
5 min 13%
more 85%

Figure 4.18 Tube/rail access: 1-, 2- and 5-minute walk band in Clissold and Bensham Manor.

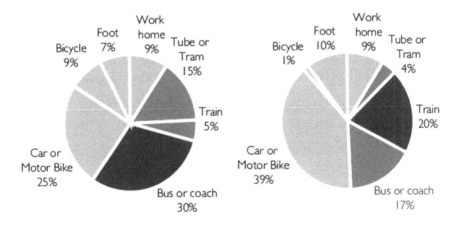

Figure 4.19 Modal split of trips to work in Clissold (*left*) and Bensham Manor (*right*).

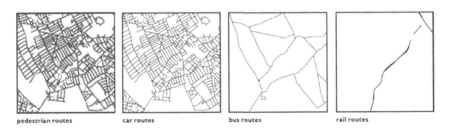

pedestrian routes   car routes   bus routes   rail routes

Figure 4.20 Transport networks, Town.

covering the entire area, however buses serve only a selected number of routes and rail lines offer only a single linear service. Personal mode-based transport always provides denser networks, but it is the absolute proximity to network routes that matters, allowing public transport to be increasingly attractive with increasing levels of density.

Constraints for accessibility resulting from higher density

Density-related constraints for accessibility are predominantly linked to transport, being the various mechanisms by which people and goods can move from one place to another. Whereas accessibility based on physical proximity increases with higher density levels, the surface space per person that can be dedicated to transport decreases. This constraint affects both, the actual process of moving people (all transport modes), as well as the requirement for parking vehicles of individual transport modes (cars, motorbikes, bicycles) and can lead to problems with transport modes with low space efficiency, particularly the private car.

Figure 4.21 shows the space requirements per person for urban transport modes such as walking, biking, public transport and car use. The inefficient use of space by the private car becomes most evident in comparison with public transport. At both speed levels indicated (30 and 50 km/h), buses require only 5%, light rail even less with about 3% of the space per person. Whereas walking is at a similar efficiency as public transport, and biking already requires more space but is still far below car use [Rode and Gipp 2001].

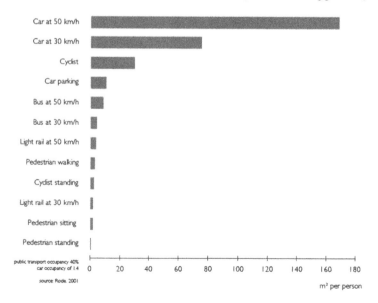

Figure 4.21 Space requirements of different transport modes.

As a result of a more comprehensive model, Figure 4.22 shows the factorised space requirements for different modal splits, taking into account the percentage of people travelling by public transport, private car and by non-motorised modes (walking and biking). This overview includes parking requirements. Here, the conflict of private car use in high-density areas can be derived from the enormous space requirement of car-dependent modal splits. The most extreme case of 100% car use in comparison to 100% public transport use shows that about 68 times the space is required for the car extreme [Rode and Gipp 2001].

Automobility constraints, particularly regarding parking, are a significant negative consideration about high density from a resident's perspective. This was confirmed by the research conducted in our five higher-density areas. The survey showed that parking availability (38%) is the second single point and the level of traffic congestion (31%) the fifth single point that most needs improving.

Looking first at residents stating that parking needs improving across the five different wards, a very clear picture emerges in relation to car density (the number of cars in a ward divided by the area). Figure 4.23 shows the percentages of people agreeing that parking needs improving for each of the wards and how this corresponds to the car density in each ward. The two elements appear to correlate with each other, leaving Town and Green Street East with the highest level of dissatisfaction with parking as well as the highest car density.

Particularly for Town, with a car density above 50 cars/ha, the qualitative interviews with local residents and actors emphasised the constraints of higher density for auto mobility.

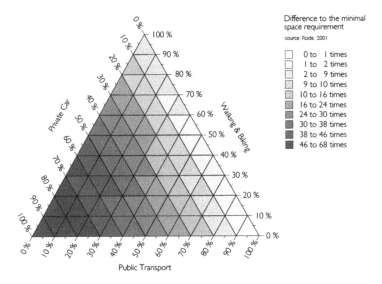

Figure 4.22 Modal split pyramid.

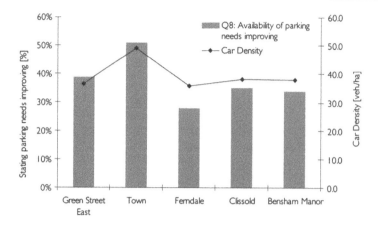

Figure 4.23 Parking problems and car density.

it's [density] negative for the reasons that I first touched on, i.e. Parking, driving around, the amount of times you drive down a street to pull in and let someone go past, because they are quite narrow streets and you have an awful lot of people who have large 4x4's! God knows why when they are in an area like Fulham. (17) (estate agent, Town)

it is too crowded, there is not the parking, most of the houses are Victorian houses, and they don't have garages so you can't park your car. ... Travelling around isn't easy because the roads are so congested although the borough does have parking restrictions and they've tried to make life a little bit simpler for residents, but it still is busy with cars. (14) (Head of School, Town)

Figure 4.24 shows the comparison between the percentage of residents thinking that levels of congestion need improving and car density. The

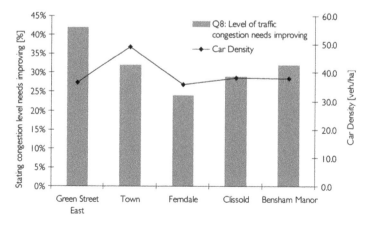

Figure 4.24 Congestion and car density.

pattern is similar, although Green Street East appears to have overproportionate congestion stress.

The high car density in Town, being the most successful dense neighbourhood of our case studies in terms of satisfaction as well as the most affluent one, raises an important issue.

Analysing the correlation of residential density and car ownership for Greater London, as shown in Figure 4.25, a clear pattern emerges. As expected, the two correlate highly in a negative way by a factor of –0.704. However, this correlation is much stronger in Outer London than in Inner London, where there is a significant number of high-density areas of above 150 pers./ha having between 250 and 400 cars/ha. These also happen to be the most affluent areas, Town being one of them.

The situation to deal with is that the more affluent neighbourhoods are, the higher the demand for automobility even if these areas are located in an urban environment of high public transport accessibility. In these cases the car does not serve for commuting (only 17% of trips to work in Town are by car) or as a guarantee for basic mobility needs. The car stands much more for the possibility to 'escape', particularly in the evening and at weekends (34% of trips unrelated to work by car), as a status symbol (high proportion of Sports Utility Vehicles as mentioned in interviews) and as a mode of transport that offers a certain convenience, safety and possibility for goods movement.

Recognising both the demand for automobility as well as the enormous problems of car densities above 40 cars/ha, successful urban neighbourhoods have to offer as much convenience and safety of car use as possible with other modes of transport, invest into alternatives to car ownership such as car sharing (90% of the time privately owned cars are not used and

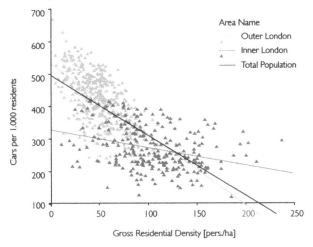

Figure 4.25 Scatter plot of car ownership and gross residential density. Significant correlation Pearson's coefficient: −0.704 for total population.

Source: Census 2001.

lead to significant parking problems) and invest in the quality of the public realm, so that it is seen as a precious public good and not as a parking lot.

Density opportunities and constraints create mobility patterns

As a result of both the constraints and the opportunities of higher densities in relation to accessibility, Figures 4.26 and 4.27 show the scatter plot of

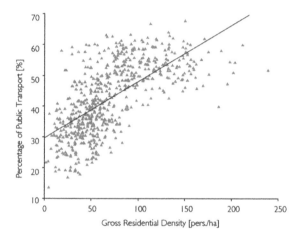

Figure 4.26 Scatter plot of trips to work by public transport and gross residential density. Significant correlation Pearson's coefficient: 0.652.

Source: Census 2001.

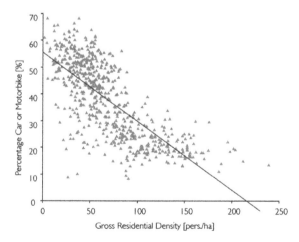

Figure 4.27 Scatter plot of trips to work by car or motorbike and gross residential density. Significant correlation Pearson's coefficient: −0.725.

Source: Census 2001.

the percentages of travel to work by public transport and by private car for all Greater London wards, and their correlation with the gross residential density. In addition to the dependence of mobility patterns on density levels, Figure 4.28 page emphasises the important influence of location and centrality for modal choices.

The emerging pattern of a positive correlation between gross residential density and public transport use, as well as a negative correlation between density and car use, is very clear. Only density levels of more than 100 pers./ha ensure that the percentages of public transport trips to work remain above 40%. Above the same density level, driving a car as the most density averse transport mode remains below 40%.

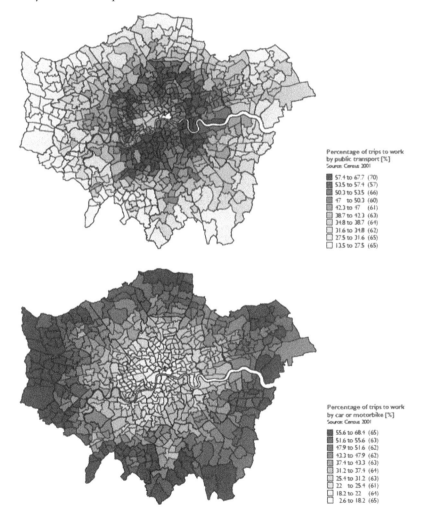

Figure 4.28 Modal shares across London by public transport (above) and car/motorbike (below).

Additional analysis showed that extremely high percentages of bus trips (above 30%) in Greater London also occur only in areas of gross residential densities of 100 pers./ha or above. This represents a density level that was identified by previous studies in order to ensure a viable bus service with bus stops at a distance of less than 500 m for 70% of residents [Urban Task Force 1999]. It remains to be said that walking and biking profit from higher densities even more than public transport. This re- emphasises the importance of the logic pair 'density and proximity'.

REFERENCES

CENSUS. 2001. Office of National Statistics. www.ons.gov.uk/census/2011census/20 11censusdata/2001censusdata (published by the Office for National Statistics)

DOT 2000: Indices of Derivation for Wards in England, 2000, Indices of Deprivation and Classifications, England, Department of Transport, Local Government and the Regions.

Häussermann, Hartmut. 2004. *Phänomenologie und Struktur städtischer Dichte. Working Paper.* Avenir Suisse and ETH Zürich, Zürich.

Land Registry. 2001. *Changes of Ownership by Dwelling Price.*

Office of the Deputy Prime Minister (ODPM). 01/08/2003. Internet: http://www.odpm. gov.uk/stellent/groups/odpm_planning/documents/page/odpm_plan_023322. hcsp#P14_261

Office of the Deputy Prime Minister (ODPM). 01/03/2004. David Cross. Project Manager for GIS and Data Systems. Planning & Land Use Statistics Division. Office of the Deputy Prime Minister, London.

Office for National Statistics (ONS). https://www.ons.gov.uk/census/2011census

Rode, Philipp and Gipp, Christoph. 2001. *Dynamic Spaces.* Technical University, Berlin.

Town and Country Planning Association (TCPA). 2003. Town and Country Planning Association Policy Statement: Residential Density.

Urban Task Force. 1999. Towards an urban renaissance. London. Final Report of the Urban Task Force. Department of the Environment, Transport and the Regions, Spon Press, London.

Chapter 5

Urban layouts, densities and transportation planning

Shirish B. Patel

CONTENTS

This chapter is concerned with how the process of urban planning should be integrated with transportation planning for the urban area. Both should be different aspects of a single seamless process sharing a common vision and shared objectives. The bulk of the chapter is concerned with the specifics of the layouts of urban localities, and how these affect densities and, therefore, ultimately travel demand.

URBAN PLANNING

Urban planning should be a three-step process:

(a) First, a declaration of principles that will drive the planning process and serve thereafter as signposts: reference markers against which each subsequent policy or project proposal is judged. Typical principles might be the following:
   i.   Provide municipal services to all income groups in the city.
   ii. All localities must be mixed-income and mixed-use localities with a variety of housing sizes to suit different income groups.
   iii. When resettlement is undertaken, and plots or pitches assigned, the period of lease will be not less than 30 years; and when such a lease comes up for renewal this must be based on a fair valuation

derived from current market prices, unless the land is needed for some more urgent public purpose.

iv. Public transport has priority over private transport: each receives funding in proportion to the number of its users.

v. Discourage the use of cars; manage traffic with a combination of improved technology, better policing and better pricing policies.

vi. Encourage preservation of the character of the city.

vii. Expand green spaces and make them accessible to all, within walking distance of where people live.

viii. The intensity of development should be variable, with more concentrated development near transit nodes.

Such principles, and their order of priority, will vary from city to city. They should be established after careful debate that involves the public. They should be reviewed and readjusted every five to ten years.

(b) The second stage of planning should be the formulation of a strategy plan for the city that looks far into the future, including, in particular, the economic future, and sets out the basic strategies for dealing with the fundamental issues of land use, transportation, water supply, sewage treatment, solid waste disposal sites, power supply and the creation and maintenance of a GIS database for the region.

An essential constituent of the strategy plan would be a parallel, companion plan for financing and recovery of investment. Without this, no part of the strategy plan proposal can even begin to be considered. This cannot be sufficiently emphasized. All plans for urban development in the past have remained just that: plans only, with no clear direction as to how they were to be implemented. In future we should expect every plan, whether at the strategy level or at the more detailed local level, to be accompanied by a detailed, companion financial plan, which would cover how capital is to be raised, where the funds for servicing loans will come from and how borrowed capital is to be repaid. Acceptance of the strategy plan should necessarily require simultaneous acceptance of the financial plan. It is only if this is done that the plan proposals will be meaningful and implementable. Pure technical plans, even if accompanied by cost implications, in the absence of clear tax and financial arrangements, often remain little more than wishful thinking.

The strategy plan should also set out the guidelines to be followed for the next level of detailed area planning, including in particular, specifying the range of desirable densities and various kinds of possible building controls. Note that transportation and land-use planning would go hand in hand with densities of occupation, an integral part of land-use planning. Mixed land use could be permitted in any zone, with restrictions if desired, but with very few fixed prescriptions at this strategic level about which particular plot of land should be used for what.

Prepared at the level of the regional planning authority, the strategic plan should be reviewed every five years. It would be sent in draft form to the next, lower, level of planning for comments and suggestions. Differences of opinion would be resolved after giving the public an opportunity to be heard. A mechanism would be in place for time-bound conflict resolution.

(c) The third and final stage would be the preparation of detailed area plans that would show detailed land-use planning and development controls. The preparation of these local area plans would be the responsibility of the various local authorities, typically acting for a locally elected municipal councillor. They would also be reviewed every five years and would cover the following:

i.   Water distribution network.
ii.   Storm drainage network, including rainwater harvesting.
iii. Sewage collection network.
iv.   Solid waste primary collection, including sorting of garbage.
v.   Location of bus stops, taxi parking ranks, public parking.
vi.   Power distribution.
vii. Road and footpath widening.
viii. Location and demarcation of public open spaces.
ix.   Location and sizing of various public amenities, including public toilets, schools, colleges, hospitals, police stations and fire stations.
x.   Definition of land uses at the detailed, local level, in consonance with the overall regional guidelines.
xi.   Development of building control regulations, which may vary from one locality to another within the local area, but which are again in consonance with the guidelines spelt out in the regional plan.
xii. For both the planning and the implementation of the above, we may involve what might be called citizens' Area Sabhas. These would be much like Mumbai's Advanced Locality Management groups, except that each Area Sabha should have a footprint coinciding with that of an election polling station (between 800 and 1,500 persons).

Here too, as in the case of the strategy plan, it is important to have a companion document that sets out the tax and financial arrangements proposed so that successful implementation can be assured.

The vexed question in the matter of detailed land-use planning is the determination of the rights of land owners in the locality. There are two problems here: one concerns the respect to be accorded to individual plot boundaries; the other is the decision as to where (on whose plot) a public facility such as a school or a public toilet should be located.

The first issue, of readjusting plot boundaries to provide more public space, should be resolved on the basis that the primary owner of all land in an urban area is the public authority, and that the ownership of individual owners is secondary and subordinate to the first. We have years of experience of town planning schemes, where an entire locality is taken up for development, a layout is prepared, and existing land owners are given plots in the new development that are smaller than their earlier plots but that are in proportion to the earlier holding. The same mechanism could apply to a situation of redevelopment where roads need to be widened or common facilities provided that did not exist earlier.

Where exactly a particular public facility, such as a school or hospital, should be located is normally open to a variety of options. Why one particular plot owner rather than another should be selected—even if they are, in principle, compensated in some way for giving up their land—is the matter to be resolved. The answer surely lies in making the giving up of land for a public purpose so attractive that the public authority will have multiple offers to choose from, and can do so following any procedure that is open to scrutiny and fair to all.

One instrument whose potential has not been fully explored is the right to construct floor space. Today that is the floor space index (FSI), which defines the ratio of buildable floor space to plot area. It is assumed that this must be some kind of constant number across the whole city, or at least across large swathes of it. Now there is no reason why ownership of land must necessarily carry with it the implicit permission to build large amounts of floor space on it. The right to build could be separate from the ownership of land. Once this principle is accepted, the right to build could, for example, arise from the surrender of land for a public purpose. The owner of the right to build is free to sell that right to someone else who owns land but cannot build on it unless he purchases that right to build from someone else. The public authority should, of course, be careful that it parts with rights to build only in such measure as the locality can sustain. These rights to build may be issued by the public authority in a variety of ways and for a variety of reasons: in exchange for land to be used for a public purpose; or sold for money to finance infrastructure; or given in exchange for the construction of middle- or low-income housing. The point is to treat the right to build as a tradeable commodity in urban areas, subject to such constraints on its use as may be important in particular localities.

There is another vexed question, and this concerns the rights of occupants of lands and buildings, who are not the owners but who happen to be in possession, either as legitimate tenants or as squatters. Mumbai still has tenants whose rents are frozen at Second World War levels. And it has half its total population living in slums—that is, in unauthorized occupation on land that belongs to someone else. Slum-dwellers in Mumbai have acquired a kind of semi-legitimacy, which does not exist, for example, in Delhi, where slum settlements are ruthlessly bulldozed regardless of how long they have been there, and regardless also of whether years ago they were officially put there in the first place by the government. In Mumbai the difference is that slum-dwellers throughout the city have been promised that if they can prove that they came to live wherever they happen to be before 1 January 1995 (the current "cut-off" date, which keeps getting regularly extended) then they can continue to live where they are. Not only that, they will be provided with free pucca accommodation, constructed without cost to them and without effort on their part, at their current location. This is achieved by inviting developers to build free housing for slum-dwellers financed through free sale in the open market of an equivalent amount of floor area, both on the same plot. Profits from the free sale are more than enough to pay for the free housing for slum-dwellers. This scheme has been shown to work. Unfortunately, it can only work for those pockets of slum-dwellings that are in localities where property prices are high, and hence the high profits realized are sufficient to pay for the free construction. Moreover, if the scheme is expected to cover the entire population of slum-dwellers, which is half of Mumbai's total, it can only work if there is a large enough number of new city residents willing to buy however many new flats, scattered throughout the city where various slums now exist.

This kind of large market for so many new high-value flats sounds unrealistic. But there is another more fundamental problem, to which we now turn our attention.

URBAN LAYOUTS AND DENSITIES

Every person living in an urban area experiences a variety of spaces. We can place these in three broad categories:

(a) Private spaces. This includes home, which means spaces private to one's family and friends, and it includes shared private spaces that one shares with one's neighbours. These may be built-up (staircases, landings) or open (the compound of one's building). The usefulness of the open space for the occupants of the building varies depending on the shape of the space, and whether parking in it is allowed or not. When you have the "island" kind of layout, where the building forms the island and the open space all around is its perimeter, if parking is allowed and the perimeter is essentially the driveway, the common open space is of little use to the residents except as a parking facility. If you have a "courtyard" type of layout, where the built form encloses a courtyard, this can form a useful common space for interaction between residents, particularly if vehicles are not allowed in the courtyard.

(b) Public spaces. These are shared with a wider public, people one does not necessarily know. Here the spaces may be one of the following:

    i. built-up spaces, for hospitals, schools, police stations, the fire station, electric substations and other common amenities

    ii. open for recreation (parks and playgrounds)

    iii. open for pedestrian circulation (footpaths)

    iv. open and reserved for bicycles (bicycle paths)

    v. open for circulation (local roads)

    vi. open for parking

(c) Arterial transport spaces. These are the transport arteries of the city, and include railway tracks and stations, expressways or arterial roads, and busways. Some roads may be partly arterial (carrying the through traffic) and partly local (including for local circulation, side parking and footpaths). For such a road, normally one would assume one lane on either side as being for local circulation. If side parking is allowed, another lane is excluded, from one or both sides as the case may be. The rest of the road width is arterial transport space.

Our interest for the moment is in the configuration of private and public spaces, and how variations in these affect densities and the working of the urban area. The relationship with transport spaces is dealt with separately later. All we note in passing is that transport capacities are related to the numbers of persons to be carried and, therefore, to densities (persons per square kilometer) in the areas served.

We should note also that urban planning and, in particular, the planning of individual urban layouts when we get down to detailed area planning, is concerned with precisely this relationship between private and public spaces, and that the various kinds of controls imposed on development on the private plots, whether by way of FSI or otherwise, are intended to control the densities in the area.

If we consider any segment of an urban area (call it a locality), of the order of a few hundred hectares in size—recall that a square kilometer is 100 hectares, so we are talking of a few square kilometers—and if we look at the layout and the built form, we can extract a variety of parameters computed from the purely physical characteristics of that area. From the range of possible parameters that apply to an area, five have been selected as being particularly significant. These are defined below:

DEFINITIONS

Buildable plot ratio: This is the ratio of the area of all the buildable plots in a locality to the total area of the locality. The plots may be used for residential, commercial, industrial or mixed use, or any other commercially exploitable construction. Excluded are plots on which public amenities are constructed, such as police stations, schools, hospitals and so on (even if such public amenities are commercially profitable). The area of these public amenity plots is included in the next category.

Public ground area (PGA) per capita: This is the area on the ground devoted to footpaths, roads, recreation spaces, school buildings and their playgrounds, hospitals, fire stations, police stations, public toilets and any other public amenities that are open to the general public.

Densities (persons/sq km, or persons/ha): This is the number of persons per square kilometre of the locality, or a hundredth of that, which is persons per hectare, a smaller and more convenient number. It will be different at night, when only the residential population of the locality is present (*nighttime densities*), and in the daytime when jobholders and visitors (including shoppers) will be added, but some of the residents may be out of the locality (*daytime densities*). Densities can be expressed as overall or gross or global densities, that is, over the entire area of the locality; or they can be densities on the buildable plot areas, sometimes called net densities. For residential localities the *nighttime density* is often expressed as dwelling units per hectare (DU/ha), because this is a parameter that can be controlled by building regulations, and this again can be DU/gross ha, or DU/buildable ha. Multiplying it by the average size of a household for any locality gives a reasonable indication of residential people density in that locality. Much of the confusion in understanding urban densities arises from this plethora of descriptive parameters.

Built-up area (BUA) per capita: This is the amount of floor space, on average, consumed per capita. It may be either *residential* or *commercial*. Both will vary according to income group: the residential BUA in Mumbai for the poor may typically be 5 sq m per capita (a family of five living in a built-up space of 25 sq m), with higher-income residents consuming perhaps 20 sq m per capita (a family of four with two servants living in a 120 sq m apartment). In New York's Community District 8 (Upper East Side, just east of Central Park) the residential BUA is 63.7 sq m per resident. In the Island City of Mumbai it is 5.3 sq m per resident. In Lajpatnagar in Delhi it is 18.3 sq m per capita. In Sunder Nagar, Kaka Nagar and Bapa Nagar (taken together) it is 98 sq m per capita.

Plot floor space index: Plot FSI is the ratio of built-up floor space on a particular plot to the area of that buildable plot. In some countries the FSI is called the FAR (floor area ratio). Both terms mean the same thing.

These five parameters are interdependent. Specify any three which are independent of each other and the other two will follow.

One caveat needs to be recorded here. This is that in the foregoing parameter of BUA per capita we are assuming that all the population in a locality is housed in built-up residential accommodation in the locality. No account is taken of any of the population resident in commercial, industrial or institutional buildings, or in slums. In Mumbai, with half the population resident in slums, and perhaps quite a few in institutional, commercial or industrial buildings, this could lead to serious distortions in understanding a locality. While all the other four parameters will not be affected, the BUA per capita could be significantly underestimated.

Let us begin by looking at the graph that relates buildable plot ratios, gross densities and PGA per capita (Figure 5.1).[1]

The x-axis is the buildable ratio: the percentage of the locality's area that is open to exploitation by individuals for residential, commercial or industrial use. The relationship between buildable ratio and gross density is linear. The surprising feature is that because PGA per capita is the determining constraint, densities fall as the buildable ratio increases. This seems counter-intuitive: one might normally expect that the greater the area there is for building on the more people you can have in the locality, but the opposite is true. This will become clearer from the diagrams that follow.

On the graph can be seen spots that correspond to particular localities in particular cities.[2] All are localities that are a few square kilometres in size—that is, each locality is large enough that it should have its proper share of schools, medical facilities and other amenities.

A, B and C wards and Charkop are in Mumbai. The Sunder Nagar spot incorporates Kaka Nagar and Bapa Nagar. Raghubir Nagar in Delhi includes Vishal Enclave.

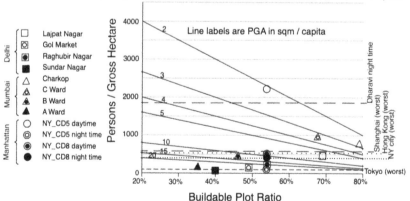

Figure 5.1 Buildable plot ratios and PGA determine gross densities.

---

[1] See Annex 1 for formulae expressing the relationships between parameters.
[2] See Annex 2 for typical details of some localities.

In addition, shown on the graph are horizontal lines that correspond to the densest localities in Tokyo (Nakano-ku), New York (Upper East Side, also called CD8), Hong Kong (Kwun Tong) and Shanghai (Nanshi). The exact spots for Tokyo's, Hong Kong's and Shanghai's worst localities will fall somewhere on the corresponding horizontal line, depending on more precise information regarding buildable ratios (this information is currently not available to the author). Dharavi is in Mumbai, reputedly Asia's largest slum, and here again, while redevelopment is in the offing, there is no information as to what the buildable plot ratio will be. Notice that Dharavi's reconstruction is being planned for night-time densities that are more than double anything so far experienced anywhere in the world.

Figures 5.2 through 5.7 will make this clear.

A legitimate question to ask ourselves at this point is: what is the PGA per capita we should assume in our planning? The PGA needed divides into three groups, the area per capita needed for (Table 5.1):

(i)   Common amenities
(ii) Open areas and recreation spaces
(iii) Roads and footpaths

In regard to Community Open Spaces (Recreation Areas) the standard for this varies widely. The international norm is 4 acres per 1,000 population, equivalent to 12 sq m per person. But it is instructive to look at different localities in different cities. A comparison will be found in Table 5.2.

Plot Area = 1 Hectare   PGA = 20 sq.m per capita
Buildable Plot Ratio 80%   BUA = 50 sq.m per capita
Footprint = 33% of plot   1 cylinder = 1 family of 5

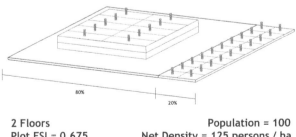

2 Floors   Population = 100
Plot FSI = 0.675   Net Density = 125 persons / ha
Global FSI = 0.5   Gross Density = 100 persons / ha

Figure 5.2 Generous 20 sq m per capita of PGA. This diagram shows a generous 20 sq m per capita of PGA, and a BUA of 50 sq m per capita, typical of a wealthy locality. We could have everyone living in G + 1 buildings, and a gross density of 100 persons per ha (G + 1 here means ground plus one upper floor).

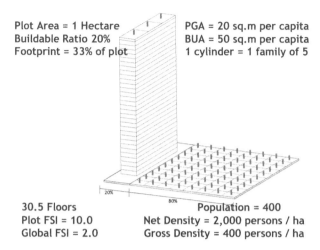

Plot Area = 1 Hectare
Buildable Ratio 20%
Footprint = 33% of plot

PGA = 20 sq.m per capita
BUA = 50 sq.m per capita
1 cylinder = 1 family of 5

30.5 Floors
Plot FSI = 10.0
Global FSI = 2.0

Population = 400
Net Density = 2,000 persons / ha
Gross Density = 400 persons / ha

Figure 5.3 Without altering PGA or BUA, more people in same area. Now, without altering PGA or BUA, that is, with no change in amenities or facilities, we could have four times as many people in the area (and a correspondingly more compact transport system), but people would be living in 30-storeyed rather than G + I buildings.

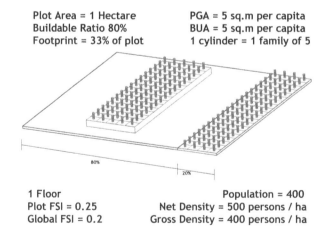

Plot Area = 1 Hectare
Buildable Ratio 80%
Footprint = 33% of plot

PGA = 5 sq.m per capita
BUA = 5 sq.m per capita
1 cylinder = 1 family of 5

1 Floor
Plot FSI = 0.25
Global FSI = 0.2

Population = 400
Net Density = 500 persons / ha
Gross Density = 400 persons / ha

Figure 5.4 Low values of poor crowded localities. Here we have PGA and BUA both 5 sq m per capita, low values typical of poor and crowded localities in Mumbai. With a buildable plot ratio of 80% everyone can be accommodated in ground floor construction, with a gross density of 400 persons/ha.

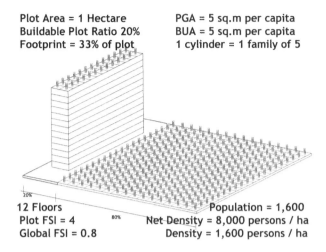

Plot Area = 1 Hectare
Buildable Plot Ratio 20%
Footprint = 33% of plot

PGA = 5 sq.m per capita
BUA = 5 sq.m per capita
1 cylinder = 1 family of 5

12 Floors
Plot FSI = 4
Global FSI = 0.8

Population = 1,600
Net Density = 8,000 persons / ha
Density = 1,600 persons / ha

Figure 5.5 Gross density of 1,600 persons/ha. Again with no change in PGA or BUA per capita, by reducing the buildable plot ratio to 20% we can accommodate four times as many people at a gross density of 1,600 persons/ha, in buildings of G + 12.

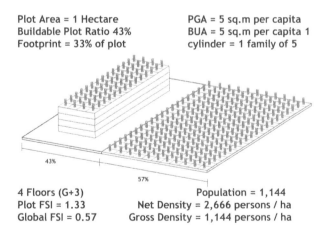

Plot Area = 1 Hectare
Buildable Plot Ratio 43%
Footprint = 33% of plot

PGA = 5 sq.m per capita
BUA = 5 sq.m per capita 1
cylinder = 1 family of 5

4 Floors (G+3)
Plot FSI = 1.33
Global FSI = 0.57

Population = 1,144
Net Density = 2,666 persons / ha
Gross Density = 1,144 persons / ha

Figure 5.6 Gross density achieved 1,144 persons. If we insist that we want walk-up accommodation, without lifts, restricted to G + 3 only, we get a reasonable buildable plot ratio of 43% for PGA and BUA both of 5 sq m per capita. The gross density achieved is 1,144 persons/ha.

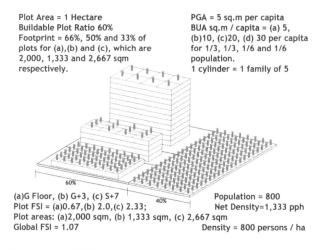

Plot Area = 1 Hectare
Buildable Plot Ratio 60%
Footprint = 66%, 50% and 33% of
plots for (a),(b) and (c), which are
2,000, 1,333 and 2,667 sqm
respectively.

PGA = 5 sq.m per capita
BUA sq.m / capita = (a) 5,
(b)10, (c)20, (d) 30 per capita
for 1/3, 1/3, 1/6 and 1/6
population.
1 cylinder = 1 family of 5

(a)G Floor, (b) G+3, (c) S+7
Plot FSI = (a)0.67,(b) 2.0,(c) 2.33;
Plot areas: (a)2,000 sqm, (b) 1,333 sqm, (c) 2,667 sqm
Global FSI = 1.07

Population = 800
Net Density=1,333 pph

Density = 800 persons / ha

Figure 5.7   Gross density 800 persons per ha. Yet another possibility is a mixed-income development, with PGA for all set at 5 sq m/capita, but BUA varying from 5 to 30 sq m/capita. The gross density is 800 persons/ha.

Table 5.1   PGA in different localities in different cities

Standards for:	National Building Code of India 2005 m^2/capita	UDPFI[a] Guidelines, August 1996 m^2/capita
Education	5.13	5.29
Health care facilities	2.07	0.84
Sociocultural facilities	0.59	0.56
Distribution services	0.04	0.04[b]
Police, civil defence and home guard	0.436	0.37
Fire	0.08	0.05
Telephone, telegraphs, postal and banking	0.1	0.1[b]
Sports activity	3.0	
Shopping	1.41	1.9
Religious	0.5	
Electrical substation	0.17	0.17[b]
Transport	0.11	0.11[b]
Cremation/Burial ground	0.13	0.13[b]
Total for common public amenities:	13.266	9.56
Open spaces (recreation):	3	13
Total for amenities and open spaces:	16.266	22.56

Notes: [a] UDPFI = Urban Development Plans Formulation and Implementation Guidelines, Ministry of Urban Affairs and Employment, Government of India, New Delhi.

[b] = guesswork.

Table 5.2 PGA in different localities in different cities

City and locality	Area (ha)	Public amenities, (sq m per capita)	Open spaces, (sq m per capita)	Roads and footpaths, (sq m per capita)	Total PGA (sq m per capita)
New York (Manhattan)					
CD-1	445.3	40.64	3.34	31.99	75.97
CD-2	402.1	7.27	0.71	16.05	24.03
CD-3	456	3.09	3.02	8.73	14.84
CD-4	542	18.42	0.83	21.09	40.34
CD-5	423.5	6.86	2.00	35.60	44.46
CD-6	354.2	4.08	0.59	8.71	13.38
CD-7	546.6	3.10	4.35	7.07	14.52
CD-8	512.9	2.98	0.44	7.46	10.88
CD-9	390.2	5.88	4.31	13.82	24.01
CD-10	363.9	3.69	1.18	12.10	16.97
CD-11	574.4	6.53	19.00	7.27	32.81
CD-12	763.5	3.62	13.38	9.83	26.83
Total (excluding Central Park)	5,775	5.83	4.94	11.42	22.19
Total (including Central Park)	6,116	5.83	7.17	11.42	24.42

Note that Manhattan has a subway railway system, and hence the pressure on roads and footpaths is very substantially reduced.

In regard to footpaths and road areas, once again a comparison of different localities in different cities is useful (see Table 5.3).

Notice that in no sub-ward of Mumbai is the road space less than 3 sq m per capita, whereas the average for the whole of Island City is over 6 sq m for roads alone. Amenities and open spaces are abominably poor and total less than 2 sq m per capita, whereas by any reasonable standard the figure should be over 16.

Turning our attention now to net densities as a function of PGA per capita and the buildable plot ratio, we see the relationship is non-linear (Figure 5.8, right-hand graph).

In Figure 5.8, note that the y-axis of densities of persons per buildable hectare can be converted to DUs/ha using the graph on the left. Mumbai has an average of 5 persons per DU, whereas New York's Community District 8 has 1.78 persons per DU.[3] When comparing localities in different cities it is useful to look at DU/ha, not just FSI.

---

[3] New York Community District 8 data from: http://www.nyc.gov/html/dcp/pdf/lucds/mn8profile.pdf.

Table 5.3   Footpath and road areas in different localities in different cities

City and locality	Area (ha)	Public amenities, (sq m per capita)	Open spaces (sq m per capita)	Roads & footpaths, (sq m per capita)	Total PGA (sq m per capita)
New Delhi					
Lajpatnagar	143				6.5
Gol Market	328				33.4
Raghubir Nagar and Vishal Enclave	375				19.91
Sunder Nagar, Kaka Nagar and Bapa Nagar	45.4				111
Mumbai (Island City)					
A-South	334.7	0.09	0.75	2.89	3.74
A-Mid	345.1	1.10	7.84	32.65	41.59
A-North	245.8	2.86	11.42	46.71	61.00
B	246.3	0.09	0.12	11.82	12.03
C	212.6	0.09	0.28	3.01	3.39
D-East	210.1	0.15	0.07	3.69	3.91
D-West	261.9	0.25	2.08	5.13	7.46
D-North	260.8	0.35	0.49	3.24	4.08
E-East	229.2	0.30	3.04	23.83	27.16
E-Mid	242.8	0.54	1.48	3.43	5.45
E-West	204.5	0.48	0.18	4.96	5.63
Mumbai (Island City)					
F/S-W	210.1	0.89	0.47	7.06	8.42
F/S-NE	157.6	0.31	0.09	9.14	9.54
F/S-SE	336.5	0.61	0.21	11.11	11.93
F/S-NW	150.6	0.63	0.78	3.55	4.97
F/N-NW	474.8	1.21	1.47	10.58	13.26
F/N-E	412.5	4.94	0.04	4.09	9.07
F/N-S	295.4	4.04	0.46	7.32	11.82
G/N-N	239.1	0.23	0.36	3.40	3.99
G/N-SE	214.9	0.21	0.14	4.79	5.13
G/N-W	277.7	0.33	0.98	5.30	6.61
G/S-N	290.7	0.48	0.28	3.50	4.26
G/S-E	287.8	0.22	0.13	4.10	4.45
G/S-W	300.6	5.54	4.17	11.05	20.77
Island City total	**6442.1**	**0.96**	**0.85**	**6.38**	**8.19**

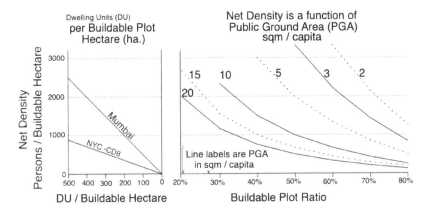

Figure 5.8   Net densities can be converted to DU/ha, using persons/DU.

The various lines on the right-hand graph are for different amounts of PGA per capita. Obviously, if we want to increase densities we have to reduce PGA—that is, cut down on the space enjoyed for parks and playgrounds, and cut down on the availability of schools and hospitals. Even if we destroy all peoples' amenities, there is one demand for ground area that cannot be reduced, and that is the ground space needed per capita for footpaths and roads. When people get out of their buildings they have to be able to move about. If the locality is served by an underground railway system then the road area might be slightly reduced, because people are circulating underground and not on the road. Also, if people do not own cars, the demand for ground space for roads and parking will be less. So let us say we need a rock-bottom minimum of 3 sq m per capita as the PGA for circulation. This is the minimum figure we observe in the worst locality in Mumbai's Island City, as seen in Table 5.2.

In practice of course we cannot have a significant population entirely without amenities. There will have to be a minimum of schools, playgrounds and hospital plots. At a very low level of service this would call for an additional 3 sq m per capita of PGA. With the space needed for circulation, this makes a total of 6 sq m PGA per capita as the limit not to be transgressed. Higher values would obviously be better. Shanghai for example has 12 sq m per capita for recreation spaces alone and claims another 12 sq m per capita for roads.

Once we have fixed the PGA per capita that we are designing for, we are on one selected curve from the range of possible curves shown on the graph. Here we notice that, if we want higher densities, we have to reduce the buildable ratio: that is, have a smaller area for plots on which building takes place, with a larger area on the ground for amenities. We need to build higher, on smaller footprints. Let us say we want to go no higher than 500 DU/ha, the highest permitted by India's National Building Code.

How high will we need to go? To answer this, let us look at the other two defining parameters mentioned earlier, built-up area per capita and FSI. The kinks seen in the curves in Figure 5.8, to the right, are because of a change in presentation in scale of the x-axis; the relationship between FSI and DU/ha is actually linear (Figure 5.9).

Different societies enjoy different amounts of built-up floor space per capita. The amount depends on affordability, a relationship between earning capacity and the cost of real estate in the city in question. In Mumbai we are close to 5 sq m per capita (25 sq m of built-up area for a family of five).

We find that to get 2,500 persons per buildable hectare on the plots (corresponding in Mumbai at five persons per household to 500 DU/ha, the maximum permissible) we need an FSI of less than 1.5. A higher FSI is meaningful only if the amount of floor space consumed per capita increases to 10 sq m, or 15. Given levels of affordability this seems unrealistic.

This apparently startling result, that we can accommodate a low-income population in a high-density settlement with an FSI below 1.5, should be confirmed with an independent calculation (Table 5.4):

We should note the limitation of the development. This is that we can house 1,000 persons per ha, which is 1 lakh per sq km. The entire area of Dharavi (Mumbai's largest slum) of 217 ha can thus hold 2.17 lakhs people, with a minimum of amenities, and not much else.

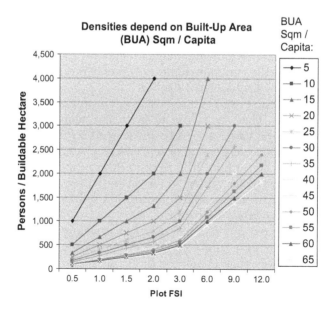

Figure 5.9 Densities depend both on BUA per capita and FSI.

Table 5.4   Low-income high-density settlement fits in G + 3

Theoretical area of total settlement	10,000 sq m
Public ground area for roads, schools, etc.	6,000 sq m
Buildable plot area @ 40%	4,000 sq m
Built-up floor space with FSI 1.5	6,000 sq m
Number of dwelling units @ 30 sq m/DU	200
Population housed	1,000
Density of dwelling units (DU/ha of buildable plot area) = 200/4000 * 10000 =	500
G + 3 buildings with 6,000 sq m floor area require 1,500 sq m plinth area, so the footprint on a 4,000 sq m plot is:	37.5%
Terrace area (which can be used for activities that require the sun) 1,500 sq m for 1,000 residents gives terrace area per resident of:	1.5 sq m

We can now reduce all the foregoing discussion to a single pair of graphs that sum up the relationship between our five fundamental parameters (see Figure 5.10).

These are complex graphs that need to be studied as a pair. They show the intricate relationship between buildable plot ratio, density, PGA, BUA and FSI. Of these five parameters, the buildable plot ratio is probably what is first determined by the urban planner as he sets out his roads and other public spaces. What is interesting is that some cities seem to have buildable plot ratios consistently around 45–50% (e.g. New York) while others show a typical range of 60–70% (e.g. Mumbai). Delhi has a particularly wide range, from 40% to 70%.

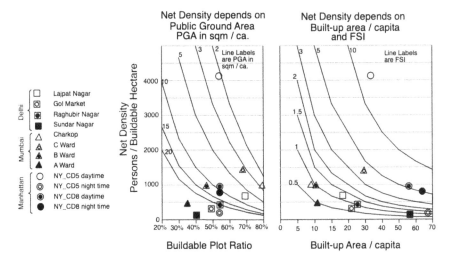

Figure 5.10   Net density depends on buildable plot ratio and PGA, and equally on BUA per capita and FSI.

The second parameter, not under the planner's control, is BUA per capita. This depends on affordability. It will vary from city to city, and from one income group to another. So in any city, for any income group, or specified mix of income groups, the BUA per capita will be known. Now if FSI is specified, as it is in so many cities' building regulations, this will determine both densities and the PGA per capita. Raising the FSI will raise the densities and lower the PGA.

Mumbai is currently embarked on a course of densification—because land is claimed to be in short supply—a course of raising FSI as a means to further densification. The current prescription is an FSI of 4 for the resettlement of slums. It will be readily apparent from the right-hand graph in Figure 5.10 that an FSI of 4 with a BUA of 5 sq m per capita leads to unthinkable densities and a PGA per capita which will be quite unworkable. There will simply not be enough space on the streets for people to move around, never mind the other public amenities that they should also enjoy. The correct answer of course is that if land is in short supply you need to focus on increasing the land supply. This can be done by, for example: building more bridges to the mainland to augment land availability; converting land that is under inappropriate use, such as salt manufacture in the heart of an urban agglomeration; cutting unnecessarily lavish allocations for use by a port whose activity should be declining; or restricting the growing of mangroves to particular ecologically sensitive areas—in all these cases it makes more sense to convert metropolitan land use to the more urgent demand of housing people and jobs, and providing both with sufficient land for the necessary essential amenities.

TRANSPORTATION PLANNING

So far we have said nothing about transportation planning, except to note in passing that transit space forms a third kind of space for urban residents, distinct from private space and public space (which we have called PGA per capita). In fact, in all the computations shown above, for any locality, the areas devoted to transit space should be taken out of the reckoning. Rail corridors and BRT on reserved corridors are all obviously transit space. For roads that partly serve local needs and are partly arterial, we can use the simple formula that one lane on each side is for local use (plus parking lanes wherever allowed) with the rest of the road considered transit space. Reserved bicycle tracks should be considered part of PGA (like footpaths), and not included in transit space.

What we need to do next is convert the transit spaces to transport capacities: how many passengers per hour per direction each can provide. For roads, this would need to take into account the current mix of vehicles on the road, and perhaps how the mix might change in future.

Transport modeling is now an enormously complex and sophisticated exercise. A metropolitan area is divided into hundreds, if not thousands, of individual pockets, each with its particular share of trip generation and trip attraction. What seems important to do is develop alternative scenarios of land use, with varying intensities of development, to see how these impact travel demand and fit within transport capacities and their possible enhancement. Equally important seems to be a stratification of trip generation and trip attraction by income groups: the travel map of the city is different for each income group. Where trips are attracted to, where they come from, and by what mode of transport they are conducted will differ from one income group to another. The land use and settlement patterns of the city will vary according to income, and transport models must account for this.

Two areas for future research seem to be of particular significance: first, how can we integrate land use and transportation planning so that both are conducted as part of a single, seamless exercise; and second, can we compare different localities around the world for their physical characteristics, and also compare them for their attractiveness as areas to live and work in, thus arriving at a range of possible values for the physical parameters that are more likely than not to deliver an enhanced quality of urban life?

ACKNOWLEDGEMENTS

The computations and graphs above of this chapter has been developed jointly with Alpa Sheth, consulting engineer, and Neha Panchal, research fellow at Kamala Raheja Vidyanidhi Institute of Architecture. The data for Mumbai has been largely obtained from Biond, a Mumbai-based geographic information system (GIS) company, with 2005 population and employment figures kindly provided by P.R.K. Murthy, chief transportation planner of Mumbai Metropolitan Region Development Authority (MMRDA). The information regarding the New Delhi areas has come from Yogita Lokhande and Nilesh Rajadhyaksha of the School of Planning and Architecture in New Delhi.

ANNEX I: PARAMETER RELATIONSHIPS

RBPA: Residential buildable plot area
CBPA: Commercial buildable plot area (includes industrial)
PGA: Public ground area
RBUA: Residential built-up area
CBUA: Commercial built-up area (includes industrial)
R: number of residents
J: number of jobs

1. Buildable plot ratio:
   a. Nighttime = RBPA/(RBPA + PGA)
   b. Daytime = (RBPA + CBPA)/(RBPA + CBPA + PGA)
2. Global densities:
   a. Nighttime = R/(RBPA + PGA)

    b. $\text{Daytime} = (R + J)/(RBPA + CBPA + PGA)$
3. Buildable plot densities:
    a. $\text{Nighttime} = R/RBPA$
    b. $\text{Daytime} = (R + J)/(RBPA + CBPA)$
4. DU/buildable ha = Buildable nighttime plot density/household size
5. FSI:
    a. $\text{Residential} = RBUA/RBPA$
    b. $\text{Commercial} = CBUA/CBPA$
6. Built-up area per capita:
    a. $\text{Residential} = RBUA/R$
    b. $\text{Commercial} = CBUA/J$

ANNEX 2: DETAILS OF PARTICULAR LOCALITIES

City	Name of locality	Area (ha)	Buildable plot ratio (%)	PGA[a] (sq m per capita)	BUA (sq m per capita)	Net density (persons/ ha)	Gross density (persons/ ha)
New York	Civic Centre, Wall Street, CD-1	445.3	41.3	75.97		187	77
	Greenwich Village, Soho, CD-2	402.1	44.4	24.03		522	232
	Lower East Side, CD-3	456	46.5	14.84		775	361
	Chelsea, Clinton, CD-4[b]	542	34.9	40.34		463	161
	Midtown, Times Square, CD-5[b]	423.5	53.8	44.46	67.3	193	104
	Murray Hill, East Midtown, CD-6[b]	354.2	48.6	13.38		791	384
	Lincoln Square, Upper West Side, CD-7[b]	546.6	44.8	14.52		848	380
	Upper East Side, CD-8[b]	512.9	53.9	10.88	63.7	785	423
	West Harlem, CD-9[b]	390.2	31.3	24.01		916	286
	Central Harlem, CD-10[b]	363.9	50.0	16.97		588	294
	East Harlem, CD-11[b]	574.4	32.7	32.81		626	205
	Washington Heights, CD-12	763.5	26.8	26.83		1020	273
	Total	5,774.6	41.2	22.19		642	265

Note: [a] Note that New York's localities are additionally served by an underground railway system (subway).

[b] All the New York localities listed surround the 341 ha of Central Park. The area of each locality includes its proportionate share of Central Park (which is less than 10% of the total).

City	Name of locality	Area (ha)	Buildable plot ratio (%)	PGA (sq m per capita)	BUA (sq m per capita)	Net density (persons/ ha)	Gross density (persons/ ha)
New Delhi	Sundernagar, Kaka Nagar and Bapa Nagar	45.4	40	111	98	141	56
	Raghubir Nagar and Vishal Enclave	375	54	19.9	25.2	342	231
	Gol Market	328	49	33.4	21.7	491	152
	Lajpat Nagar	143	69	6.5	18.3	788	469
Mumbai Island City	A-South	334.7	88.9	3.7	6.28	334.5	297.4
	A-Mid	345.1	35.2	41.6	29.06	441.7	155.7
	A-North	245.8	24.5	61.0	25.08	511.1	125.4
	B	246.3	46.1	12.0	10.78	973.2	448.3
	C	212.6	67.7	3.4	11.25	1,408.1	953.4
	D-East	210.1	69.6	3.9	11.79	1,119.3	778.7
	D-West	261.9	76.2	7.5	29.72	418.1	318.7
	D-North	260.8	78.4	4.1	19.17	673.7	528.3
	E-East	229.2	71.7	27.2	21.84	145.5	104.3
	E-Mid	242.8	61.7	5.5	7.43	1,137.3	702.0
	E-West	204.5	52.7	5.6	7.63	1,601.6	844.9
	F/S-W	210.1	59.8	8.4	9.53	977.7	584.4
	F/S-NE	157.6	55.8	9.5	5.78	829.9	463.2
	F/S-SE	336.5	74.0	11.9	4.42	294.5	217.9
	F/S-NW	150.6	71.4	5.0	7.73	805.5	575.3
	F/N-NW	474.8	62.8	13.3	19.64	446.1	280.3
	F/N-E	412.5	62.8	9.1	5.27	653.0	410.1
	F/N-S	295.4	49.5	11.8	5.22	861.7	426.9
	G/N-N	239.1	71.4	4.0	2.43	1,003.1	716.5
	G/N-SE	214.9	66.7	5.1	4.32	2,090.8	1,395.6
	G/N-W	277.7	61.9	6.6	10.57	929.3	575.7
	G/S-N	290.7	78.8	4.3	8.69	630.3	496.9
	G/S-E	287.8	75.0	4.5	5.94	874.0	655.6
	G/S-W	300.6	38.6	20.8	6.34	765.6	295.6
	Island City	6,442.1	62.6	8.2	9.62	765.8	479.1

New York (Manhattan) Community District 5 (Midtown, Times Square, Herald Square, Midtown South):

Area map	Particulars:

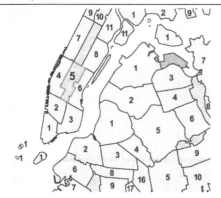

2000 population: 44,028 residents + 906,770 workers
Total land area: 1,046.4 acres = 423.5 ha (+ 38.97 ha of Central Park)
Buildable plot area: 28,709,300 sq ft = 266.72 ha
Buildable residential & mixed residential/ commercial lot area: 13.7%
Buildable commercial/office/industrial/vacant/ miscellaneous lot area: 70.7%
Built-up area residential: 31,904,870 sq ft = 296.4 ha
Built-up area commercial: 309,070,796 sq ft = 2,871.4 ha

Detailed map	Computations:

PGA buildable = $(100 - 84.4)/100$ *266.72 = 41.61 ha
PGA open = $(423.5 - 266.72 + 38.97) = 195.75$ ha
PGA total/capita (residential) = $(41.61 + 195.75)/44,028 * 10,000 = 53.91$ sq m
PGA total/capita (residential + jobs) = $(41.61 + 195.75)/(44,028 + 906,770) * 10,000 = 2.5$ sq m
BUA residential/capita = $296.4 * 10,000/44,028 = 67.32$ sq m
BUA (residential + commercial)/capita = $(296.4 + 2,871.4) * 10,000/(44,028 + 906,770) = 33.31$ sq m
Buildable ratio (residential) = $(13.7\% * 266.72)/(13.7\% * 266.72 + 41.61 + 195.75) = 13.34\%$
Buildable ratio (residential + commercial) = $(84.4\% * 266.72)/(84.4\% * 266.72 + 41.61 + 195.75) = 48.68\%$
Residential FSI = $296.4/(13.7\% * 266.72) = 8.11$
Commercial FSI = $2,871.4/(70.7\% * 266.72) = 15.23$
Residential net density = $44,028/(0.137 * 266.72) = 1,205$ persons/buildable residential ha
Residential gross density = $44,028/(0.137 * 266.72 + 41.61 + 195.75) = 160.7$ persons/ gross ha
Daytime net density = $(44,028 + 906,770)/(0.844 * 266.72) = 4,243.8$ persons/buildable ha
Daytime Gross Density = $(44,028 + 906,770)/(423.5 + 38.97) = 2,055.9$ persons/gross ha

Sources: http://www.nyc.gov/html/dcp/pdf/lucds/mn5profile.pdf.
Note: Email from Shampa Chanda, New York transportation planner, 2.12.04

Chapter 6

Coming to terms with the complexity of Indian urbanism

A. G. Krishna Menon

CONTENTS

INTRODUCTION

The correlation between urbanization and economic development is a well-established and universal phenomenon. It is therefore predictable that in India the dramatic growth of the economy will fuel massive urbanization. Even as urban planners try to cope with what already exists they will have to deal with far greater numbers in future. The current level of urbanization is less than 30% and it can easily double within the lifetime of most architects and urban planners who are practicing today. But the issues of urbanization are not just a matter of numbers, but of the increasing complexity that is defining its characteristics. In the context of finite urban space and constrained economic resources, on the one hand, and, the process of socio-cultural transformation taking place in our society, on the other, the order of complexity is such that even the well-intentioned responses of architects and urban planners become the source of problems confronting societies.

A foretaste of the kinds of problems confronting urban planners is provided in a recent report published by the Asian Coalition for Housing Rights (ACHR). It is based on the research they undertook in 2003, in eight Asian cities, to identify the process of socioeconomic, physical and institutional change that took place in those cities since the late 1980s on account of structural adjustment, the World Trade Organization (WTO) regime and the dominance of the culture and institutions of globalization in development policies at the national level. The report brought out several disturbing trends that are adversely affecting the lives of many, but particularly the disadvantaged groups in these Asian cities.[1]

As the global economic network strengthens, and the Indian economy becomes more integrated with it, the ACHR report identifies the very real prospects of increasing fragmentation of cities creating segregated zones of rich and poor. This will threaten the social, cultural and political fabric of our society, and, ultimately, its stability. Cities in India have, of course, always evolved along the lines of caste and class, but the contemporary patterns of change identified in the report involve a new scale of spatial segregation and a more complex combination of social and economic factors – one that urban planning, as a profession and discipline, has not yet begun to comprehend, let alone learned to handle. This professional inability is rooted in the history of the profession.

URBAN PLANNING: PAST AND PRESENT

Town planning has been practiced in India for millennia. There is archaeological evidence of well-planned towns from the period of the Indus Valley civilization. There are several treatises on town planning that attest to the existence of theoretical discourse on the subject in ancient and medieval India. From many literary sources, we also know that medieval towns were well structured, though in their contemporary form it would be difficult to discern the ordering principles that governed their layout. But with colonization, town planning in India abandoned its indigenous roots and adopted instead European principles of town planning.

In effect, the dialectic process of resolving local urban issues was replaced by unquestioningly using foreign models of planning. This process was understandable during the colonial period, but, unfortunately, even after Independence, it continues to define official town planning policy. In this chapter I seek to examine this problem by analyzing the origins of modern town planning in India in order to throw light on our current predicament. Pari passu, I hope to explain the theme of this volume the problems confronting transportation planning.

After the rebellion of 1857, there was a concerted effort by the colonial government to rebuild Indian towns on terms they could "understand". The government realized that they had come a hair's breadth of losing their empire in the cities because their organic morphologies made them difficult to control. Therefore, after they had regained control, they set about transforming Indian cities into the form and image with which they were familiar. Veena Talwar Oldenburg has compellingly documented this process in her book, *Making of Colonial Lucknow, 1856–1877*.[2] The birth of modern town planning in India is rooted in these imperialist intents.

Arindam Dutta points out in his book, *The Bureaucracy of Beauty*, "It is because the origins of modernity in the colony are inexplicably tied up with the ends of imperialism that its outlines operate as a historical teleology in reverse: First the institutions and then the 'enlightenment'".[3] This process characterizes the evolution of town planning in India. It begins to explain why the process of making a plan or "solution" is already determined before understanding the "problem". Consequently, the government, to establish its authority, has to deploy the police powers of the state to "enlighten" society and ensure that the "solution" is duly fitted to the "problem". This narrative of institution-building can begin to explain the process of "sealings and demolitions" that was experienced by the citizens of Delhi in the recent past.

To institutionalize the "modern" process of planning, the Public Works Department published *The Handbook on Town Planning* in 1876, which contained the guidelines for undertaking town planning projects in India.[4] These guidelines were produced at a time when the imperial government had begun systematic efforts to ensure civic health and hygiene in their own cities at home, so they merely transferred the models they developed for their cities to the colonies. Nevertheless, their interest in incorporating new ideas was evident even then, because the *Handbook* was updated eight times in the 70 years before Independence, with each successive edition including the advances in urban planning practices that were taking place in the United Kingdom. In the 60 years after Independence, however, the Indian Public Works Department (PWD), the heirs to this legacy, republished this book only twice, but without attempting any major changes to its contents. Thus the original references to Ebenezer Howard's Garden City and the New Town plans developed in England between the two World Wars were retained as models to be used in India. It is possible that the irrelevance of its contents may have resulted in the fact that no one referred to it any longer, and the book itself is not traceable, but its neglect after Independence illustrates two debilitating characteristics of urban planning in India: one, the unquestioning adoption of foreign models to serve local purposes; and two, the inability or unwillingness among professionals to engage intellectually with the emerging complexity of Indian urbanism.

It is therefore not surprising that urban planning, as a profession and discipline, has not been able to understand the nature of Indian urbanism – both as a sociocultural construct and as a technical entity. Instead it has always tried to "discipline and punish" its natural propensities rather than mediate it, let alone facilitate the process. It is therefore possible to construe from this brief analysis that urban planning, as it is practiced in India, cannot come to terms with the contemporary complexities of Indian urbanism, including the issues of transport planning.

But this is not a new revelation. As far back as 1924, file notes in government records identified the problem of local developments not following the "Master Plan", formulated by Edwin Lutyens. In fact, Stephen Legg concludes, in his study of the development of New Delhi, that the "would-be panopticism of the imperial city became impossible to regulate from the very beginning".[5] A report of 1939 reproduced the 1914 sanctioned layout of New Delhi, a coloured map of stark, clearly defined functional zones that contrasted with the map of the actual layout of New Delhi in 1938, which presented a dazzling array of mixed land uses and "haphazard developments". Even then the problem was not identified at the level of supply and demand, or of understanding local cultural propensities, but as one of a lack of strict regulation. Thus the main recommendations of the report centred around regulating housing standards, designs and skylines. In the evolving government policies after Independence, stricter enforcement became the leitmotif to solve the problems of "illegal" development. Such entrenched epistemological assumptions of policy makers and urban planners are now routine and must in themselves be considered a problem in order to chart new directions to plan our cities.

Another dimension of the problem of urban planning in India that Legg points out is that it deals with the city as an abstract entity, and not as a living organism. There is no intellectual or moral commitment to the positions planners take – only a feeling of victimhood because people do not follow their prescriptive plans. Perhaps this indifference to the consequences of policies began after Independence. Jawaharlal Nehru attempted to combine the existing capitalist system with a Soviet-inspired plan of socialist, industrial, state-planned development as a "scientific" and supposedly "objective domain". It necessitated planning by experts to protect national interests: it was assumed that local benefits would follow. The aim was to rule "from a distance" through targeting supposedly autonomous epistemological objects such as an economy, society or population. A similar condition prevailed in urban planning.

The point to be emphasized is that this process of planning "from a distance" can be viewed as one where urban planners have been trying to fit a cube into a round hole. Unless they appreciate the implications of this paradox they will not be able to come to terms with the complexity of Indian urbanism. For this to happen urban planners will have to take into account the ground realities while formulating masterplans for cities.

IMAGES AND GROUND REALITIES

There is therefore a strong case for changing the way town planners conceive the city. Their imagination needs to be stimulated and developed right from the time they begin their studies. But academic institutions are mostly concerned with passing on received knowledge and practical experience for minimally informed and vocational ends. There are no serious studies of Indian cities based on conscious hypotheses. This is because there is little or no history of town planning thought in India. Under the circumstances, town planners have not formed image(s) of the city other than those derived from cultural, social and economic societies different from theirs.[6] There is a need to think of the city in indigenous terms that incorporate the

culturally plural, socially evolving and economically constrained characteristics of Indian society. Such an enterprise has been long overdue: there
is an urgent need to bury the mentality represented by the PWD *Handbook
on Town Planning* in order to "decolonize" the concept of the city.

Nevertheless, it is necessary to take into account the positive aspects of
the radical and lasting changes in Indian society and culture produced by
colonization. It was unlike any previous period in Indian history, as the
British brought with them new technology, institutional knowledge, beliefs
and values. The changes included what Daniel Lerner called a "disquiet-
ing positivist spirit", touching public institutions as well as private aspira-
tions … People came to see the social future as something that could be
manipulated rather than ordained, and their personal prospects in terms of
achievement rather than heritage.[7]

Thus, when I speak of "decolonization" and getting rid of instruments
like the PWD *Handbook*, I do not mean to imply that everything that
changed in Indian society on account of colonization should be exorcised.
Through careful consideration in each instance, there is a need to distinguish between instruments that were relevant in a pre-colonial and colonial
context and what is relevant today.

In the case of town planning, there is a need to distinguish between the
way town planners conceived the city then, and the way cities ought to be
conceived today. As the profession of town planning was established during the colonial period, and town planners saw the Indian city as outsiders
looking into a culturally alien situation. They viewed what existed in negative terms and considered the benchmark from England in positive terms
and as the normative model. Oldenburg has shown in her study on colonial Lucknow that the supposedly "neutral" political, social, economic and
physical changes made in the city after the rebellion of 1857, brought about
constitutionally aggressive and purposive control of the population, and
that these changes continue to be accepted as desirable strategies by contemporary Indian urban planners in lieu of more appropriate ones based
on the local imperatives of societal welfare.[8] Changing the existing town
planning paradigm is now an inescapable conclusion, and changing the
way town planners conceive the city an important beginning.

RESTRUCTURING KNOWLEDGE AND CONCEPTS

To accomplish this change will require the kind of research that is currently
not taking place in academic institutions in India. We will need to restruc-
ture the curricula – but where do we begin? From ancient times there exist
theological treaties and commentaries that few town planners understand
or would find relevant to the contemporary context. Pre-colonial literature
on cities and towns is also not useful because they are characterized by
"biographical" writings, that is, the biography of a town.[9] It is necessary

to distinguish between biographies and serious studies based on conscious hypotheses. It is only when the latter objective is addressed that we will get to understand the nature of the Indian city, its institutions, its social-psychological make-up, and the ethos of urban classes. To date, there have been few attempts to understand the complexity of the urban condition in India (with the major exception of the Report of the National Commission on Urbanization[10]), resulting in a restriction of the boundaries of the town planner's knowledge.

To begin with it is necessary to recognize the singularity of our urban condition. It derives from the fact that our society has widely plural characteristics, temporally, culturally and economically. Such a condition does not exist in other societies, old or new, and while we may gain insights through cross-cultural references on certain issues, it should not eliminate the need to do our own homework. The complexity of the situation can be gauged by the reality that in town planning terms, not one, but several disparate circumstances need to be reconciled simultaneously: neat suburban developments with homogenous population *and* the persistence of the heterogenous, "'chaotic", traditional settlements; the city of the "haves" *and* the city of the "have-nots"; Lutyens' baroque city *and* the *qasba*; the automobile *and* the bicycle; and the like. There are few models to conceptualize such heterogeneity anywhere, so town planners in India will have to become increasingly self-referential. Notwithstanding the problems and pitfalls inherent in this perspective, there are promising clues that can be fruitfully explored.

One promising area of enquiry lies within the field of urban conservation. With the establishment of the Indian National Trust for Art and Cultural Heritage (INTACH) in 1984, there was a focused interest in the conservation of our built heritage. It soon became clear that the museum-like conservation practices of the West could not be the model for India, and that we would have to view our heritage in *developmental* terms. This was the underlying premise of the "heritage zone" concept proposed by INTACH. There is reason to believe that this concept has a wider application.[11]

The INTACH projects demonstrated that a study of traditional settlements with a view to *developing* them offers a rich mine of information that could be rewarding and could be exploited by town planners in other situations too. Such an exercise forces the modern town planner to abandon the *Handbook* approach and focus on the particularities of the built-up areas within towns. These pre-colonial parts of the city have, so far, been both academically and physically neglected and are, therefore, beyond the consciousness of the modern town planner in India. However, the traditionally evolved settlements are the repository of culturally embedded methods and devices that are useful for planning contemporary cities.[12]

Having worked on conservation-oriented development proposals for such historic cities as Varanasi, Ujjain, Old Bhubaneswar and Chanderi,[13] I can state with a certain degree of confidence that these apparently intractable urban problems can be resolved. Half a century ago, Patrick Geddes had convincingly demonstrated that this was possible in a manner that the modern town planner needs to re-examine. Geddes looked at the city as an organic system that was amenable to a carefully structured process of healing and the encouragement of natural growth. His approach was context-specific: it was both effective and satisfactory.[14]

The urban conservation projects with which I was involved were necessarily also context-specific. They forced us to look more closely at the origins of the problems in order to find an appropriate response rather than rearranging them to fit a predetermined order derived from other cultures. This did not mean that there were no overarching objectives in our projects: there were. We had broad objectives such as improved quality of life at the local level, sustainable development at the level of society, and "people-first" approaches to problem-solving in general. Of course, there were other issues, too, such as accepting the evidence of tradition as the norm, but suffice it to say that this was a different way of conceiving the city than had hitherto been attempted by town planners. What emerged in this process was what Robert Venturi calls a "both-and" environment.[15]

Not surprisingly, one begins to realize that in terms of settlement density, social heterogeneity and economic mix these traditional cities are examples of an urban character not seen in other parts of the city developed by modern town planners. While they have real problems, such as infrastructural and other forms of deprivation, as urban typology they are both satisfying and appropriate models for imagining the Indian city. It is in this context that the recently initiated Jawaharlal Nehru National Urban Renewal Mission (JNNURM) must be viewed in positive terms because it has the potential to achieve this objective.

It is interesting to note that even in America there is some debate beginning to surface on the need for a "non-Euclidian mode of planning"[16] to meet the future needs of society. This "locally oriented, socially informed, politically astute, results-oriented planning" describes the kind of planning proposed to be undertaken in the JNNURM programme. INTACH's heritage zone concept was also predicated on this proposition.

When these views are presented to town planners at the helm of affairs, their characteristic response is that it would necessitate entirely new ways of planning, and that such radical changes in town planning practice would not be possible: legally, practically and politically. But, ironically, it is this very kind of radical change that underpins JNNURM. The JNNURM puts into practice the principles enunciated in the 73rd and 74th amendments to the Constitution. It will usher in far-reaching changes in the way town planning is practiced in India.[17] But the field of enquiry is still wide open. I suggest that here is an opportunity that the town planner really must seize: the JNNURM should finally motivate the profession to bury the mentality of the PWD *Handbook on Town Planning* once and for all and reconceptualize the city in more appropriate terms. Perhaps it is still possible to come to terms with the complexity of Indian urbanism.

REFERENCES

1. David Satterthwaite, *Understanding Asian Cities*, Asian Coalition for Housing Rights, Bangkok, 2003.
2. Veena Telwar Oldenburg, *Making of Colonial Lucknow 1856–1877*, Princeton University Press, Princeton, 1984.
3. Arindam Dutta, *The Bureaucracy of Beauty, Design in the Age of its Global Reproducibility*, Routledge, New York, 2007.
4. B. T. Talim, *PWD Handbook, Chapter II, Town Planning*, 10th Edition, Government Printing and Stationary, Bombay, 1976.
5. Stephen Legg, Post-colonial developmentalities: from the Delhi Improvement Trust to the Delhi/Development Authority, in Saraswati Raju, M. Satish Kumar and Stuart Corbridge (eds), *Colonial and. Post-Colonial Geographies of India*, Sage Publications, New Delhi, 2006.
6. It is often assumed that it is difficult to understand a Third World metropolis. The truth is that this has rarely been attempted. A recent attempt by a World Bank team can serve as a model: Rakesh Mohan, *Understanding the Developing Metropolis: Lesson from the City Study of Bogota and Cali, Colombia*, The World Bank/Oxford University Press, Oxford, New York, 1994. The ACHR report cited above is another example.
7. Daniel Lerner, *The Passing of Traditional Society*, Free Press, Glencoe, IL, 1958, pp 45–9.
8. Veena Talwar Oldenburg, op. cit. The titles of the chapters of her book make a telling commentary on the contemporary imperatives of town planners: *The City Must Be Safe* (where she discusses demolitions and the building of segregated enclaves); *The City Must Be Orderly* (the role of the Police and the Municipal Committee); *The City Must Be Clean* (sanitation and building bye-laws which changed the morphology of the city, and hence its social livability); *The City Must Pay* (the concept of penal tax); and *The City Must Be Loyal* (the making of a loyal elite). Do contemporary town planners think any differently?.
9. S. C. Misra, Urban history in India: possibilities and perspectives, in Indu Banga (ed.), *The City in Indian History, Urban Demography, Society and Politics*, Manohar Publications, Delhi, 1981.
10. Ministry of Urban Development. The National Commission on Urbanization – Report – 2 Volumes, Government of India, 1988.
11. A. G. K. Menon, Conservation in India: A Search for Directions, *Architecture + Design*, 1989, pp. 22–27.
12. A. G. K. Menon, *Cultural Identity and Urban Development*, INTACH, New Delhi, 1989.
13. INTACH, Unpublished Reports, Available from INTACH, 71, Lodi Estate, New Delhi.
14. Jacqueline Tyrwhitt (ed.), *Patrick Geddes in India*, Lund Humphries, London, 1947. For an interesting commentary on the issue of context-specificity, see A. K. Ramanujam, Is there an Indian way of thinking? An informal essay, *Contribution to Indian Sociology* (New Series) Vol. 23, No. 1, 1989, pp 41–58.

15. Robert Venturi, *Complexity and Contradiction in Architecture*, The Museum of Modern Art, New York, 1966. Indeed, one is tempted to paraphrase Venturi's often quoted comment on the Main Street: "*Traditional Cities* are almost always alright!".

16. See John Friedmann, Toward a non-Euclidian mode of planning, *Journal of the American Planning Association*, Vol. 59, No. 4, Autumn 1993, pp 482–5. Also commentaries on pp 485–6 and *APA Journal*, Vol. 60, No. 3, Summer 1994, pp 372–9.

17. Several seminars and workshops have been organized by the Institute of Town Planners, India, recently, See D. S. Meshram, Changes needed in State, Municipal and Town Planning Acts consistent with the Constitution 74th Amendment Act, 1992. An overview, *Journal of ITPI*, New Delhi, Vol. 12, No. 3, March 1994, p 157.

Chapter 7

Urban mobility

Is anyone in charge?

K. C. Sivaramakrishna

CONTENTS

INTRODUCTION

In discussing urban mobility, apart from demographic and spatial factors such as population increase, density, pattern of land use, and employment locations, there are other important aspects, such as per capita trip generation, transit volume, modal split, vehicular increase, etc. Mobility in cities is also to be considered for both passenger and goods movement. Even in the best of circumstances, when resources are not a constraint, urban mobility is determined more as a result of public policy and the response of public agencies to market forces. This chapter seeks to emphasize the fact that, in most Indian cities, the poor state of urban mobility is due more to public policy failures and the fact that at the city level no one appears to be in charge of urban transport.

India's total population according to the 2001 Census is 1.27 billion. The urban share of this population is about 285 million, which is about 27.78%. This is about 21% of Asia's and 10% of the world's urban population. However, these simple percentages do not convey an adequate picture of the country's urban scene. Much of the urban growth is concentrated in large cities. There are 35 cities of more than 1 million people, which have a total population of about 108 million people, representing nearly 40% of the urban population. These million plus or "metropolitan cities", as they

Table 7.1 Growth of megacities

Sl. no.	Metropolitan cities/UAS	Population			Exponential growth rate	
		1981	1991	2001	1981–91	1991–2001
1	Greater Mumbai	8,243,405	12,596,243	16,368,084	4.22	2.62
2	Kolkata	9,194,018	11,021,918	13,216,546	1.72	1.82
3	Delhi	5,729,283	8,419,084	12,791,458	3.80	4.18
4	Chennai	4,289,347	5,421,985	6,424,624	2.23	1.70
5	Bangalore	2,921,751	4,130,288	5,686,844	3.36	3.20
6	Hyderabad	2,545,836	4,344,437	5,533,640	5.20	2.42

are usually referred to, have been growing steadily. At the time of the 1991 Census there were only 24: now there are 35.

Those with a population of more than 5 million are commonly referred to as "megacities". There are six of these at present, namely Greater Bombay (renamed recently as Mumbai), Calcutta (renamed as Kolkata), Delhi, Madras (renamed as Chennai), Bangalore, and Hyderabad. The populations of these six megacities since 1981 are given in Table 7.1. It should be noted that these are according to the area defined by the Census, which may not cover the full area of metropolitan agglomerations. Although on average the population of these megacities increased by about 1.89 times during 1981–2001, the number of registered vehicles went up nearly eight times.

VEHICULAR GROWTH

The growth of motorized vehicles in most of the metropolitan cities has been dramatically high. In the 15-year period between 1980 and 1995 the number of vehicles increased by 334% in Delhi, 229% in Calcutta, 116% in Bombay, and 371% in Bangalore. In comparison, the population increase itself has been 67%, 25%, 29% and 60% respectively. In metropolitan cities as a whole, the number of motorized vehicles is 40 per thousand population. However, in some cities this ratio is much higher. Table 7.2 indicates the current levels as well as growth in recent years.

The two-wheeler phenomenon is unique to Indian cities. The Lambretta-type scooter with the two-stroke engine, discarded in European cities by the 1970s, gained a massive entry into Asian cities, in particular, India. From a small number of 27,000 for the whole country in 1951, the production of these vehicles jumped to 1 million by 1976 and 2.3 million in 1996. In 2000, the number of scooters, including an increasing number of motorcycles produced in the country, reached 3.4 million. The Finance Ministry's Economic Survey for 2006–07 indicates production during 2005–06 was 7.6 million. About 500,000 were exported during the year leaving the

Table 7.2 Details of vehicles registered in major cities

Urban agglomeration/district	Two-wheelers	Autos/ tempos	Cars cabs	Buses	Goods carriages	Tractors	Total
Bangalore							
01.04.1985	195,210	123,75	58,971	3,812	12,217	5,881	288,466
01.04.1990	4158,54	15,754	85,037	4,243	18,298	6,555	545,741
01.04.1995	594,639	34,335	120,103	6,454	24,625	14,220	794,376
01.04.2002	1,183,752	64,520	259,001	10,077	49,037	30,171	1,596,558
Chennai							
01.04.1985	122,123	6,115	55,529	2,945	12,337	798	199,847
01.04.1990	338,486	5,580	113,783	3,657	28,917	1,871	492,294
01.04.1995	565,451	20,849	148,896	4,317	23,475	1,460	764,448
01.04.2002	988,630	44,771	250,080	4,541	31,459	6,202	1,325,683
Delhi							
01.04.1985	579,064	30,017	166,263	13,522	52,370		841,236
01.04.1990	1,113,236	58,934	354,810	17,844	92,778		1,637,602
01.04.1995	1,617,732	7,4981	588,309	26,202	125,071		2,32,295
01.04.2002	2,265,955	86,985	989,522	47,578	161,650		3,551,690
Greater Mumbai							
01.04.1985	167,338	32,351	236,186	22,506	46,840	3,427	508,648
01.04.1990	305,099	41,814	302,122	10,878	75,405	3,623	738,941
01.04.1995	434,802	101,597	313,537	16,291	83,517	6,070	955,814
01.04.2002	787,527	212,862	547,224	20,718	124,718	8,215	1,701,264
Kolkata							
01.04.1985	160,556	4,968	177,736	15,736	62,514	10,375	431,885
01.04.1990	217,304	7,100	219,079	18,330	75,083	11,480	548,376
01.04.1995	294,110	10,146	270,039	21,352	90,179	12,703	698,529
01.04.2002	467,756	27,003	380,079	28,923	105,687	28,003	1,037,451

balance on Indian roads, mostly in the metropolitan cities. Delhi is not only the capital, but it is also the scooter capital of the country. Out of a total of 3.55 million vehicles registered between April 1985 and April 2002, nearly 2.27 million are two-wheelers. Until recently these vehicles used two-stroke engines. It is only in the past three years that industry has shifted to four-stroke engine technology. As indicated in Table 7.2, in the other cities of Bangalore, Madras, Bombay, and Calcutta, the predominance of the two-wheeler is the same. It is important to understand that scooters and motorcycles represent such a fast changeover from public to private transport because of the personal convenience and ease of purchasing. The capital cost, on average, is about US $4,000 and the market is such that there is quick, easy financing.

The two-wheeler boom is also a part of the rapid proliferation of private motor vehicles. While the production of motor vehicles in general has been increasing in recent years the manufacture of private cars has gone up significantly. Compared to 2.34 lakhs vehicles in 1998–99, production had more than doubled in 2003 at 7.81 lakhs vehicles. According to the Economic Survey, it crossed 1.04 million in 2005–06. Though the automobile industry claims a rise in the export figures from 161,000 in 2004–05 to 170,000 during the following year, the bulk of the production is sold within the country. The Ministry of Road Transport Research Wing brings out the *Road Transport Year Book*, which contains some very useful data. Table 7.3 indicates the production of different types of motor vehicles as well as their sales over a six-year period. In a 15-year period, between 1985 and 2000, the number of vehicles increased by 371% in Bangalore, 334% in Delhi, 229% in Kolkata, and 116% in Mumbai. The population increases in the cities certainly do not, in themselves, warrant this. The increase is more the result of rising incomes and the consumption patterns. In every metropolitan city, the increase in motor vehicles, particularly non-transport vehicles, has been alarming, and even rampant in a few cities. For instance, in the one-year period, 2002–03, Delhi added 86,624 cars, which works out to 237 per day. It is reported during the calendar year 2005, the figure crossed 90,000. The number of two-wheelers increased by 1.63 lakhs or 447 per day. Chennai was not to be outdone, with an addition of 86,260 cars (236 per day) and 2.54 lakhs two-wheelers, which works out to nearly 700 per day. Compared to this the additions to the numbers of buses are modest, to say the least.

The automobile industry is regarded as a key segment of the economy, with extensive forward and backward linkages with other segments of the economy. Since 2001–02, installed capacity has been growing at a compounded rate of 16% annually. However, very little attention is given to the impact on the cities. The automobile industry itself is quite content to demand that road space in the cities should be expanded, traffic facilities, such as flyovers, improved, traffic managed better, slow-moving vehicles banned, etc. Some of these measures may well be desirable and need not be decried merely because the automobile industry is seeking them. But there

Table 7.3 Production of motor vehicles in India (in 000)

Category	1998–99	2002–03	2003–04	% change of 2003–04 Over 2002–03
1	2	3	4	5
M&HCVs	80	120,502	166,102	37.84
LCVs	55	83,195	109,122	31.16
Total commercial vehicles	135	203,697	275,224	35.11
Cars	391	557,410	781,764	40.25
Multi-utility vehicles	113	165,920	206,776	24.62
Scooters	1,315	848,434	935,319	10.24
Motorcycles	1,387	3,876,175	4,355,137	12.36
Mopeds	672	351,612	334,494	−4.87
Total two-wheelers	3,374	5,076,221	5,624,950	10.81
Three-wheelers	209	276,719	340,729	23.13
Grand total	4,223,469	6,279,967	7,229,443 (rounded)	

Source: *Road Transport Year Book 2003–2004.*

Notes: M&HCVs = medium and heavy commercial vehicles. LCVs = light commercial vehicles.

During 2004–05 and 2005–06, the production of two-wheelers rose to 6,527,000 and 7,600,000; and of passenger cars 842,000 respectively.

are limits to road space expansion. Delhi is considered to be one of the better endowed cities in this regard, with about 16% of the land space. Other cities are not that fortunate. But even in Delhi, with its passion for flyovers (over 20 have been contracted in the past three/four years), there is little relief to buses or pedestrians. Whatever benefits accrue from these flyovers, they are only temporary and rapidly used up by the increase in the number of vehicles. In most Indian cities, road space available ranges from 6% to 12%. Governments then seek to "create" space through flyovers and elevated highways but, as noted by environmentalists, "ever increasing cars fill the ever increasing space".

Congestion, therefore, is widespread and vehicle speeds on average are between 8 to 12 km per hour in the megacities. The congestion factor is also reflected in accidents and fatalities. For the country as a whole, about 400,000 road accidents were reported during the year 2000, resulting in about 79,000 fatalities and 400,000 persons being injured. The proportion of this in metropolitan cities is high. Figures available for 1998 indicate that, in the five megacities considered in this chapter, accidents numbered about 60,000 with 4,355 fatalities. In addition to the loss of lives, the economic costs of congestion are very high – as is repeatedly confirmed by international experience.

PASSENGER MOBILITY

Given the vehicular growth and congestion we can thus consider the situation regarding the mobility of passengers. It is estimated that per capita trip generation in megacities is about 0.8 to 1.1. The broad break-up of the modal split of the transit volume for four megacities is indicated in Table 7.4.

Public transport modes in these megacities have had different histories. In the cases of Bombay and Calcutta, where growth has been mainly linear, the suburban railways carry a significant volume of goods and people. The suburban railways are not the same as intercity commuter trains, though they are operated by the same railway companies sharing the track infrastructure. In both Calcutta and Bombay, the suburban trains are electric multiple units, transporting passengers from the suburban through the metropolitan area and terminating in the city centre. Stations are located all along the route and, therefore, passengers make use of these trains within the metropolitan area as well. In the case of Madras, suburban train services were developed on the metre gauge in three radial directions. In all these three cities the trains account for nearly one-third or more of the transit volume. In the case of Delhi, buses are the principal means of transit. In Madras and Calcutta, they account for important shares of the transit as well. Because of the limited road space, the need to serve different localities and need for flexibility in operating schedule, minibuses have also been introduced in these megacities. However, the volume they carry is still limited.

DISTORTIONS IN PRICING

In considering pricing in urban transport there are several aspects that need to be taken into account. One is the cost of building and maintaining road space. The second is the allocation of that road space for various modes of

Table 7.4 Modal split (percentages)

Transit volume (million trips)	Madras 3.95	Bangalore 3.33	Delhi 11.95	Calcutta 9.3
Two-wheeler	7	50	30	–
Three-wheeler	3.8	20	1.7	11
Car	1.5	20	28.3	16
Bus	37.9	5	36.2	–
Suburban train	4.1	–	–	18
Train	–	–	–	48
Walk	29.5	–	1.6	2
Metro	–	–	–	2
Others	14.2	5	2.8	3

transport, giving preference to public transport assuming that to be the polity. To what extent such use is paid for, if at all, is another aspect. Fourth, there is the cost of acquiring vehicles and the taxes associated with that. Fifth, there are the operating costs of different transport modes, including fuel costs. Sixth is the direct cost of traffic management. While these are all quantifiable and measurable, there are also significant social costs arising from congestion, accidents, air pollution, etc. In most countries data about these costs are difficult to get. The Indian cities are no exception.

The bulk of the cost in building and maintaining road space is borne by public authorities. In most cases, this will be the respective state governments. In the case of Delhi, however, being the nation's capital, it receives a significant amount of central government funding. For the 11th Plan period (2007–12), the outlay proposed is Rs.885,700 lakhs on roads and bridges and Rs.205,456 lakhs on transport. Delhi's expectation is that the bulk of the outlay on transport will be the mass rapid transport system (MRTS), amounting to Rs.161,732 lakhs. During 2006–07, Delhi spent about Rs.93,622 lakhs on roads and bridges and less than half of that amount, i.e. Rs.38,664 lakhs, on road transport. Here again the bulk of the expenditure is on the MRTS: Rs.29,998 lakhs. Though the state government is a party to the project, the bulk of the funds are received through a complex financial package, funded significantly by Japanese aid and supported by the central government. While figures for other states are not available, it can be safely assumed that the expenditure on road and bridges will be at least three or four times more than the expenditure on transport itself.

Regarding the priority for public transport when allocating road space, although this is the declared policy, in most Indian cities the implementation is only on a token basis. This is partly because of the severe competition for road space by proliferating private transport and also because of the poor enforcement of any priority measures. As for user charges directly levied for accessing road space, this is limited to just a few bridges and tollways in some other cities. Tolls on some highways are seen more widely but within cities this is not the case. For example, in all of Delhi, there is only one bridge across the Yamuna where a toll is collected, and this bridge was privately built. Elsewhere in the city, or in any of the other megacities, no charges are levied at all for use of the numerous flyovers and elevated roadways that have been constructed at considerable public expense in recent years. There is not one city that has attempted to levy a "cess", or a surcharge, on vehicles operating in the congested parts of the city. Even parking fees are a recent introduction and the rates continue to be low.

Motor vehicle tax, or road tax as it is popularly known, is one levy collected by all the states, either on a one-time-only basis or annually. The revenue realized for some states in India is given in Table 7.5. It will be seen that, except in Andhra, Gujarat, Karnataka, Maharashtra, and Orissa, these taxes do not account for the bulk of the revenue. In the case of Delhi, out of Rs.3,815 lakhs as road tax revenue, the contribution of private

Table 7.5 Revenue realized from motor vehicles taxes, fees, etc. for 2002–03 (Rs. in lakhs)

States/UTs	Total	Motor vehicle tax	Commercial vehicle and other fees	Passenger tax	Goods tax	Fines
States						
Andhra Pradesh	91,869.00	16,736.00	6,566.00	21,823.00	35,704.00	9,040.00
Gujarat	81,180.88	51,091.82	7,320.69	1,019.36	21.72	21,272.29
Karnataka	69,749.71	42,715.87	5,188.36	...	17,617.75	4,227.73[a]
Madhya Pradesh	43,542.00	7,576.00	8,190.00	13,406.00	14,370.00	...
Maharashtra	105,853.00	90,244.00	140.00	11,573.00	...	18.22
Punjab	44,370.55	8,549.91	5,450.37	26,675.00	1,945.88	1,749.30
Rajasthan	64,605.27	7,579.42	20,537.93	16,114.62	16,360.68	4,012.62
Tamil Nadu	73,934.97	27,241.28	17,356.81	16,884.71	11,492.94	1,959.23
Uttar Pradesh	85,093.47	24,475.03	23,372.64	19,135.56	16,034.24	2,076.00
West Bengal	25,072.00	...	...	...	...	...
UTs						
Delhi	3,815.49	470.17	743.20	1,171.19	1,195.85	235.08

Source: *Road Transport Year Book 2003–2004.*

Notes: [a] Rs.55.52 lakhs shown as other also included. ... = not indicated

motor vehicle tax is no more than Rs.470 lakhs, whereas bus passenger tax came to three times that figure, i.e. Rs.1,071 lakhs. The rates and patterns of motor vehicle tax for private vehicles as well as for buses are archaic, arbitrary, and indicate no particular rationale of revenue yield or equity.

Private sector money also goes into the production and distribution of mainly private vehicles. For instance the cumulative number of buses manufactured in the country as a whole, as of the year 2000, is less than 600,000, compared with a total of 48 million other vehicles. Between 1999 and 2000, the cumulative number of buses increased only by 19,000, compared to half a million cars and two million two-wheelers. It is also common knowledge that financing a private car purchase is one of the easiest and quickest transactions in India. Compared to this funding, financing at reasonable interest rates remains a time-consuming and difficult exercise for taxis and buses. To aggravate matters, vehicle taxation has been unimaginative and favours private transport vehicles rather than buses. In many states of India, private vehicles pay a one-time-only tax ranging from US $25 to US $100 at the time of registration, which bears little relation to vehicle cost. On the other hand buses pay a tax calculated on the basis of number of passengers, including standing passengers.

The impact of such taxes on bus passengers usually tends to be regressive compared to the case of private vehicles (as illustrated by Table 7.6). A World Bank study of 2004 confirms that the total tax burden per vehicle kilometre is Rs.5.69 in the case of buses, Rs.2.39 in the case of cars, Rs.1.03 for taxis, and Rs.0.44 for two-wheelers.

MASS TRANSIT

The need for mass transit and the scope for its adoption in the major cites of the country has been discussed for more than three decades. Feasibility studies and project reports are far too numerous to list. However, concrete action has been inversely proportional to the amount of debate and reports. The first effort to build a dedicated rail-based mass transit system was taken in Calcutta in the 1970s and was intended to be partly underground and partly on the surface. The initial project envisaged a 30 km system expected to carry about 20% of the transit volume, but what is now operating is a 13 km line used for about 10% of its capacity and 2% of the total transit volume. Underfunded from the beginning, and delayed at every stage, the metro rail in Calcutta reached an operational stage in 1978. In the case of Chennai (Madras), one of the suburban train lines has been extended to the city in stages as some kind of transit facility. In the case of Bombay, a World Bank-financed transit improvement project under implementation is focusing on improving the frequency of the suburban train services, road improvements, and improvements in the bus system. A mass transit project of two corridors, one about 12 km long and another 38 km long has been

Table 7.6 Motor vehicle taxes and incidence (per person per trip) for private and public vehicles

	Private		*Public buses*
	Two-wheelers	*Cars*	
Gujarat	Unladen weight (ULW) of 50 –	5% of cost	>9 seats = Rs.840 p.a. + Rs.72 per additional seat
Incidence	Rs.0.17	Rs.1.25	Rs.0.04
Maharashtra	7% of cost	4% of cost	Rs.71 per seat p.a. + 17.5% of fare collected
Incidence	Rs. 0.41	Rs.1.00	N/A
Karnataka	7% of cost or Rs.2,500, whichever higher	7% of cost or Rs.18,000, whichever higher	Rs.2,000 per seat p.a.
Incidence	Rs.0.42	Rs.1.75	Rs.0.83
Tamil Nadu	Rs.200 p.a.	Rs.800 p.a.	Rs.1,100 per seat p.a. + 10% surcharge[a]
Incidence	Rs.0.17	Rs.0.33	Rs.0.50
West Bengal	Rs.1,800	Rs.1,200 p.a.	>33 seats = Rs.2,475 p.a. + Rs.40 per additional seat
Incidence	Rs.0.30	Rs.0.50	Rs.0.03
Uttar Pradesh	Rs.1,500 + Rs. 150 p.a.	2.5% of cost or Rs.5,000, whichever higher, + Rs.500 p.a.	>36 seats = Rs.4,460 p.a. + Rs.180 per additional seat
Incidence	Rs.0.38	Rs.0.83	Rs.0.06
Delhi	Rs.1,220	Rs.4,880	>19 seats = Rs.1915 p.a. + Rs.280 per additional seat
Incidence	Rs.0.20	Rs.0.41	Rs.0.09
Haryana	Rs.150	Rs.1,500	Rs.550 per seat p.a. up to max. of Rs.35,000
Incidence	Rs.0.03	Rs.0.13	Rs. 0.23

Source: *Road Transport Yearbook 2003–2004*.

Notes: Unless mentioned otherwise, all taxes are one-time. [a] Formula applies only to buses on town service.

Assumptions

Approx. no. of trips per year per two-wheeler = 1,200.

All cars are petrol-run. Average price of car = Rs.3 lakh. Average unladen weight: 800–1,200 kg.

Approx. no. of trips per year per car = 1,200.

All buses are diesel-run. Average passengers per bus = 50. Approx. no. of trips per year per bus = 2,400.

Approx. life of a two-wheeler/car = 5 years.

proposed, and bids for a private–public partnership arrangement have been initiated. Bangalore, after years of debate, has finally commenced work on a MRTS system consisting of an east–west corridor of 18 km and another north–south corridor of 15 km. The project, being executed by a joint venture company, is expecting to complete the first 7 km of line by 2011.

The first phase of the Delhi metro system covers a length of about 65 km, out of which 12.5 km is underground and the rest on the surface or partly elevated, at an estimated cost of US $2.2 million. Delhi metro is undoubtedly expensive. But thanks to significant Japanese aid and the political clout that Delhi can exercise in mobilizing funds, the first phase has been completed in a little over seven years. In a country otherwise plagued by notorious delays in implementing public projects, the metro construction agency in Delhi has been a singular exception. This success has also prompted the neighbouring jurisdictions of Delhi, such as Noida and Gurgaon, to seek an extension of the metro system to their areas. The Commonwealth Games, held in Delhi in 2010, gave a further boost to the metro. However, attention to alternative forms of transit is still limited.

After several years of persistence, a beginning is being made in regard to dedicated busways under the JNNURM (Jawaharlal Nehru National Urban Renewal Mission). To provide better public transport and ease congestion, proposals for a bus rapid transit system (BRTS) have been approved for Ahmedabad, Bhopal, Indore, and Pune under JNNURM. The approved schemes, covering a total length of more that 156 km, are estimated to cost a total of Rs.1,408 crore, of which central assistance is around Rs.670 crore. Considering the low cost, ease of implementation, wide area coverage, and overall sustainability, a lot of cities are coming up with BRTS proposals to be funded under JNURM.

INSTITUTIONAL FRAGMENTATION

That the current situation in most Indian cities is fragmented will be stating the obvious. Roads and their maintenance are usually the responsibility of the public works departments of the state (provincial), while for some roads it is a city government responsibility. The registration of motor vehicles, licensing, and vehicle taxation are handled by regulatory authorities or departments of the state. Regulation of traffic and penalties for road violations are invariably handled by the police, which, in Indian cities, do not come under city government control. Traffic engineering, including signals, is shared between city governments and police. The operation of transport vehicles, such as buses, trains, or tramcars, is the responsibility of individual utility companies, whether owned privately or by the state. The production and sale of automobiles is in the hands of the private sector and there are no limits about the number of vehicles that can be sold. Taxation on production, including excise duties, is managed by the central government, while registration and other user fees, if any, are state-level responsibilities. Demand management hardly figures in the terminology of urban transport policies.

It is frequently lamented that the multiplicity of agencies is a principal reason for the unresolved problems of urban transport. In one sense this lament begs the question. By definition, any metropolitan area is a collection of polities. Tasks in any such area are bound to be multiple and highly varied. While they intersect with each other, it is not feasible to entrust all these multiple tasks to a single organization. Even that would not ensure that the multiple tasks would interconnect or integrate with each other. Form has to follow function, and it is necessary to understand the nature of the functions involved. In urban transport they are many. Here is a partial list:

- Land use and transport planning
- Planning and building of the circulation system, i.e. the roads, railways, etc.
- Choice of transport modes including non-motorized transport, their procurement, and operation
- Technology issues
- Routing and tariff fixing; intermodal coordination
- Traffic management
- Fuel choice and environmental safeguards
- Financing options, mobilizing funds, and allocation of public funds for transport

The list could be expanded, and could be rearranged in many different ways, but it is obvious that all these are important and critical elements of urban transport.

The National Urban Transport Policy, adopted by the government of India in 2006, recommends the setting up of a unified metropolitan transport authority in each million-plus city to facilitate better coordination in the planning and implementation of urban transport systems. It has been suggested that such an authority should not be an operator for any transport facility but should function as a coordinator among various operators. Its role would be more a regulatory one and possibly that of a provider of common facilities. Funding for urban transport projects under the JNNURM is contingent on the setting up of the Unified Metropolitan Transport Authority (UMTA). It is possible that this will indeed happen. Even if the cities are not totally convinced about the usefulness of such a body, it is likely that the lure of funds will prompt them to establish these bodies without a clear idea about the design. The chances are that the role of these UMTAs will be limited to intermodal operations, routing, and possibly tariff fixing but not much beyond that.

We need to consider which functions are critical and have serious policy implications and what is the minimum organizational framework that is needed to have these functions performed. In metropolitan cities comprising multiple municipal jurisdictions, the tasks became even more complicated. Some Asian cities, such as Tokyo or Seoul (or Manila or Bangkok, using special purpose coordination agencies) have made some modest progress through metropolitan governmental structures. In the United States, where the private automobile continues to be king, freeways abound, and various forms of public transport have to carefully steer a course through a maze of local authority regulations, limited fare revenues, and varied public choices, the Federal Transit Administration (FTA) of the Department of Transportation has long struggled to establish a coordinated approach, at least for planning. The Intermodal Surface Transportation Equity Act (ISTEA), enacted in 1991, makes available significant federal assistance to metropolitan areas, provided the local authorities can prepare and subscribe to coordinated surface transport planning. This act has since been replaced by the Safe, Accountable, Flexible Efficient Transportation Equity Act – Legacy for Users of 2005 (SAFETEA-LU). Under this act, the FTA supports locally planned and operated public mass transit systems throughout the United States. Though the multiplicity of local authorities is a veritable maze, in most American cities, including New York and Los Angles, the fact remains that the transport systems are somehow kept going. The Council of Local Governments, known as COGs, where different local bodies representing cities and counties share a platform of coordination, has shown some modest success. In the Washington, DC metropolitan area, for example, there are about a dozen local government bodies, which somehow cobble together the funds and decisions needed to keep the metro system running. Weighted voting rights for different COG members, depending on their populations represented, is a special feature.

The integration of land use and transportation has been a favoured item in the curricula of planning schools, but the fact remains that contemporary realities are not factored in. For example, cities that have concentrated employment in the centre require high-capacity transit. Bombay is an apt example of this, where the concentration of employment on the island and the long commute from the extended suburbs both justified and reinforced this pattern. But where activities are scattered and the spatial configuration is polynodal, a different kind of transport is called for. The concept of self-contained neighbourhoods favoured in the British new towns, and followed in some Indian cities, requires that the size of each neighbourhood is small and that daily needs are within walking distance. This is not the case in many Indian cities. We have also had problems with the concept of so-called non-conforming uses, since mixed land uses are a part of the traditional Indian city's caps. The most recently published Delhi Master Plan for 2021 talks about synergy between land use and transport, and flags integrated multimodal public transport systems to reduce dependence on personalized vehicles as an objective, but there is little evidence of how this is to be achieved. Yet the Delhi Plan may be considered a little better than plans for other cities, where the transport implications of land use do not receive much attention. Even in the case of the new towns of India, built after Independence, be they the steel towns of Durgapur, Bhilai, Ranchi, Rourkela, or Bokaro, or administrative capitals such as Chandigarh, Bhubaneshwar, or Gandhinagar, the land use/transport aspect has not received much attention. The distances between centres of activity and the residential zones do not facilitate ease of movement. Land use/transport integration thus continues to be a failure on the planning front.

Given the fact that India has to depend very much on fuel imports, the demand management of proliferating private transport should have been regarded as an organizing principle of public policy. Yet, as explained in the previous paragraphs, this is an aspect that has been largely ignored. Irrespective of fuel supply or fuel pricing, car sales have continued to grow and they crossed the 1 million mark in just 11 months in 2006. Sports utility vehicles, which is another name for fuel guzzlers, grew at 7.5%. The World Energy Outlook, published by the International Energy Agency, foresees significant increases in oil demand in Asia. In 2003, India spent US $15 billion on oil imports, which was about 3% of the GDP. The growth rate for vehicles in the West has been around 5% per year in recent years, but in Asia it is 15% to 30% per year. Financial incentives continue to be provided for the automotive industry, but these are loaded more in favour of private vehicles than public transport.

Mention has been made about taxation and user charges. This is yet another area where distortions continue. Setting aside social costs, even direct costs associated with road use and other infrastructure facilities by private vehicles are not recovered. This is yet another area of policy failure.

The approach to non-motorized forms of transport has been erratic, to say the least. There is a strong mindset that non-motorized vehicles are a lower order of transport and should therefore be either banned altogether or relocated to the peripheries of neighbourhoods and cities. Nonetheless the demand for their services is clearly manifest, as is evidenced by the very large number of non-motorized vehicles continuing to operate in different parts of large and medium cities, notwithstanding the repressive measures taken against them from time to time. In regard to motorized modes itself, the debate on technology choice continues to be unabated but diffused.

WHERE DO WE GO FROM HERE?

It is obvious that for urban transport, planning at the metropolitan level is a first requirement. The mandate for this task can be derived from the 74th Constitutional Amendment, which prescribes a Metropolitan Planning Committee (MPC) for agglomerations with a population of more than one million and comprising two or more municipal jurisdictions. The Constitution also requires that this committee should be composed principally of representatives of the various urban and rural local bodies situated in the area. The terms of reference for the MPC are indicated in the Constitution itself. The integrated development of the infrastructure, coordinated spatial planning of the area, and environmental conservation are important aspects of the draft Development Plan to be prepared by the MPC. However, notwithstanding the constitutional prescription, only one state, West Bengal, has set up such a committee for the Kolkata metropolitan area. The Metropolitan Development Authority, set up by the state government in 1971, now functions as the technical secretariat for the MPC, which has been in existence for about five years. The Committee has been reviewing and adopting various sectoral plans for the metropolitan area including the transport plan.

The setting up of the Unified Metropolitan Transport Authority, as suggested in the National Urban Transport Policy, is also needed but the chances are its tasks will be limited to operational matters such as intermodal operation, route planning, and tariff fixing, etc. But even this simple prescription has not been implemented. The JNNURM is advocating the establishment of the UMTAs while funding urban transport projects. Hopefully the need to get central funds may prompt action since logic alone has not worked.

Transport planning and implementation require both aggregation and disaggregation. A metropolitan system of expressways and mass transit will not work if at the sub-metropolitan level connectors and feeders are not provided. The functioning of these facilities also depend on the neighbourhoods and shopping precincts, how they are planned or have grown, and the extent of connectivity needed. Current national and state policies, and market forces, will inevitably result in vehicular penetration and intrusion in every neighbourhood. The issue is how we can accommodate the voices of the neighbourhood cities in a metropolitan area.

Here again we have a constitutional prescription. At the minimum, for every city with a population of three lakhs or more a two-tier structure has been envisaged. One is the corporation or municipal council, composed of representatives elected from territorial constituencies. The Twelfth Schedule to the Constitution specifically lists urban planning, including town planning, as the very first item in the list and requires state government, by law, to endow the corporations and municipalities with the powers and the authority necessary to carry out the responsibilities contained in the schedule. Planning is thus clearly recognized as a responsibility of the urban local bodies. In metropolitan areas, the development plan to be prepared by the MPC is to be based on the plans prepared by the local bodies constituting the metropolitan area. Unfortunately most of the states have given short shrift to the constitutional provisions. Invoking some colonial prescription of inherent and superior powers, state governments interfere or interpose their departments in various levels of planning, enforcement, and interpretation. There are numerous cases in the high courts and the Supreme Court contesting the scope and validity of these intrusions. In essence the dispute is neither technical nor legal, but political, and about how the growing financial power and clout that urban land development and public works entail are to be distributed. It is to be recognized that the extensive debates on technology choices, alignments of roadways, or selection and design of flyovers are significantly influenced by this essential reality.

The Constitution also provides that, for every municipal or corporation ward that is the territorial constituency, a ward committee should be set up. This committee is to be chaired by the councillor, who is elected from that constituency, and it is left to be decided by the state government. Where the ward size is small, it is also open to the state government to group two or more wards into a committee. These ward committees are required by the Constitution to be set up for every city that has more than three lakhs population. Additionally it is left to the discretion of the state governments to provide for other committees, such as zonal committees, as yet another tier between the ward and the city council. Kolkata has had such an intermediate tier for a long time, called boroughs. In Mumbai, territorial constituencies or electoral wards are grouped into administrative wards or zones. In Delhi, Hyderabad, and Chennai, ward committees have been formed by grouping the wards.

Though the Constitution has provided this basic structure, the implementation of these provisions has been minimal and dilatory. The problems are partly administrative and financial, because very few cities have systems where the city budget can be disaggregated at the ward level or the provision of services monitored and supervised below the city level. But these are not insurmountable problems. There is also no restriction or impediment to empowering the ward committees to perform some oversight functions. Apart from discussing citywide plans for transport and suggesting changes, the ward committees can also take a proactive role in measures for parking and traffic-calming in residential neighbourhoods or shopping precincts, thereby reducing the needless conflicts between transport and people, and the potential damage to life and property that such conflict causes.

This chapter began with the question "Urban mobility, is anyone in charge?". This question is, however, part of the larger question, s "Who Is in charge of the city". The 74th Constitutional Amendment, which was preceded by extensive debates across the country, was also considered at length by a multi-party select committee of parliament. It was also ratified by the states. By far, this is the best available prescription for designing a city's governance in a decentralized and participative manner. In spite of the Constitutional prescription, implementation has been uneven at best and minimal at worst. In a matter like urban transport, which is so vital to the economic performance of the city and quality of urban life, policy failures persist, programme distortions continue, public funding is iniquitous, and performance is non-accountable. If these trends persist, the future of the Indian city is bleak indeed.

Chapter 8

Railroading the rules

Transport, government,
and stakeholders

Dunu Roy

CONTENTS

This chapter attempts to define and analyse the relationships that exist within the institutions of governance, the various sections of the population and their sectoral interests, and the transport policies that flow from such social interactions. It is based on a reading of the situation in the national capital of Delhi, although many of the trends are also visible in other cities and towns in India. Since the Hazards Centre is a technical support agency responding to community needs, the chapter is necessarily limited by the activities and experiences of the Centre with concerned and vulnerable social groups who have little access to information and research inputs.

GOVERNANCE MECHANISMS

The national capital of Delhi represents a complex system of governance, since many "stakeholders" find a foothold in the decision-making process and there are implicit conflicts and overlapping areas of governance among them (Figure 8.1).

There are, first, the *elected* institutions, which are supposed to represent the people of the capital, as well as the nation. Chief among them is the Government of India (GOI), presided over by the legislatures of parliament. It is also the largest owner of land in the country and the power of "eminent domain" over all resources resides within it. The Government of

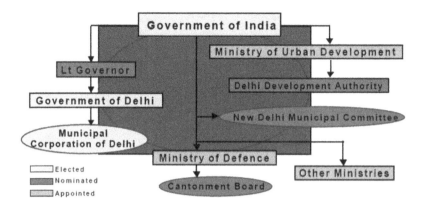

Figure 8.1 Delhi's governance structure.

the National Capital Territory of Delhi (GNCTD) represents the citizens of the capital through the Delhi Assembly, but its powers are restricted to the provision of civic and welfare services. Here too, there is another body of elected representatives in the Municipal Corporation of Delhi (MCD) that is responsible for the actual provision of amenities.

There are four other *nominated* bodies, which often exercise more powers than the elected ones. The Lieutenant Governor (LG) of Delhi, for instance, is the nominated representative of the GOI, who has to approve of the decisions of the GNCTD. He also presides over the Delhi Development Authority (DDA), which is the planning body for the city. Sharing the burden of providing urban services with the MCD are the New Delhi Municipal Council (NDMC) and the Cantonment Board (CB).

Effective executive power remains with the *bureaucracies* within the ministries. Thus, the Ministry of Urban Development is the parent body of the DDA, while the Ministry of Defence supervises the work of the CB. Other ministries, such as the Railway Ministry, the Human Resource Development Ministry, the Information and Broadcasting Ministry, the Water Resources Ministry, the Forest and Environment Ministry, and the Road Transport Ministry, also own significant areas of land in the city and can exercise considerable clout in decision making when it affects their respective constituencies.

There has also been a sharing and shifting of power bases within the various institutions of governance over the past six decades. Thus, the Ministry of Rehabilitation was set up by the GOI in 1948 to set up three camps to temporarily house the five lakh refugee population that came streaming into the city from Pakistan. In the same year the GOI also took over the transport services from the Gwalior and Northern India Transport Company to set up the Delhi Transport Services (DTS). Two years later the Ministry of Rehabilitation had completed the task of settling one lakh refugees in new houses along the ring road, while the remainder had occupied (97.5% illegally) the one lakh abandoned houses of refugees fleeing in the opposite direction, and the GOI set up the Delhi Transport Authority (DTA) to replace the DTS.

There was a partial (and temporary) transition to decentralisation in 1952 when the first Delhi Assembly was set up under the GOI. But in 1955 there was a jaundice epidemic, largely as a consequence of the establishment of the refugee colonies five years earlier, that left 700 people dead in the city. Consequently, the GOI set up the DDA in 1956 for the proper planning of the city and it was given authority over the slums under the Slum Areas Act. In 1957, Delhi was declared a Union Territory under the direct control of the LG, while the MCD and NDMC Acts were also passed to create those two bodies. The next year witnessed the historic sweepers' strike, which marked the steady deterioration of civic services, and the DTA was changed to the Delhi Transport Undertaking (DTU). The DDA also finalised its slum resettlement policy in 1958, providing for 80 square yard plots under a 99-year lease. In 1962, the First Master Plan was notified by DDA.

In 1966, the Delhi Administration Act was passed, once again constituting a Delhi Metropolitan Council under the LG. The next year the DDA arbitrarily reduced the size of the resettlement plots from 80 to 40 square yards. And, even though 1969 marked the passage of a Declaration of Social Development, in 1970 the Union Cabinet decided the plot size would be further reduced to 25 square yards, and also that the plots would be at the periphery of the city at a rent of Rs 8 per month. In 1972, the DDA decided it would give a flat of 18 square metres instead of a plot, and the payment charges were hiked to Rs 18,000. The Delhi Transport Corporation (DTC) was thus established to match the on-going commercialisation of services.

Faced with an acute housing shortage in the city during the preceding years, many families had settled at the periphery in what were called "unauthorised" colonies. Of these colonies, 567 were regularised at no cost in 1975; the Urban Land Ceiling Act was passed in 1976 outlawing the ownership of more than 500 square meters of land; and then, in 1976, under the cloak of the declaration of National Emergency, 1.5 lakh families were moved out of the city and resettled in 44 colonies at the periphery. This was also the prelude to the Asiad Games of 1980, when the formal planning process was suspended and about 10 lakh labourers came into the city for the construction work for the Games.

Under different governments, the Metropolitan Council was dissolved in 1980, revived again in 1983, and then subsumed under the National Capital Region Board in 1985. The year 1987 marked the decision of the GOI to provide basic public services, but the next year 1,500 died of a cholera epidemic in the resettlement colonies, so a three-pronged Slum Strategy was adopted in 1990. The Second Master Plan was finally notified in 1991, ten years behind schedule – under the pretext that it was merely a "modification" of the First Plan. And the elected GNCTD was given the responsibility for providing basic services in 1992, while the DDA transferred its tasks of relocation to the MCD.

In 1994, the Ward Committees were constituted to comply with the decentralisation provisions of the 74th Amendment, but these committees merely consisted of all the councillors of the notified ward. In the same year, the DDA dreamt up the ambitious Yamuna Channelisation project, estimated at Rs 1,800 crores, of which 60% was coming from joint ventures, and making available almost 10,000 hectares of land for commercial development. The same vision of the reborn city marked the filing of a public interest litigation (PIL) in 1995, asking for the closure of 168 hazardous industrial units in the city on the grounds that they were polluting the river. Eventually, this case resulted in the closure of 2,245 "polluting" units and, still later in 2000, over 100,000 "non-conforming" units. This phase also marked the intervention of the judiciary in governance.

In 1996, the Rail India Technical and Economic Service (RITES) also prepared the first plans for the modified Metro, while the GNCTD took over the DTC. In the same year another PIL was filed, asking for the removal of waste from the city, and this eventually culminated, in 2000, in court directions to remove the slums and the judicial characterisation of slum-dwellers asking for their rights to resettlement as "pickpockets". Yet another PIL was filed in1998, against growing air pollution in the city and this time the diesel buses were targeted as being the main culprits. By 2002, orders had been passed for the conversion of all buses to compressed natural gas (CNG). The Metro construction had begun a year earlier, and the Delhi Vidyut Board was privatised in the same year.

The year 2002 was not so fortunate for other residents of the city though. Under pressure from court orders and the reigning concept of "nationalism", the GOI announced an Action Plan to deport the supposedly "lakhs" of Bangladeshis in the city. The High Court pronounced a ban on begging because seven of those unfortunates had died in a beggar's home. The next year, the High Court also ordered the removal of 35,000 slum families from the banks of the river on the grounds that they were polluting the river; and the Supreme Court gave a verdict confining hawkers and vendors to certain zones. Three years later it also ordered the sealing of shops all over the city and the removal of cycle rickshaws from Chandni Chowk on grounds of "congestion".

The elected and executive institutions have tried to counter these moves of the judiciary by issuing notifications and passing legislation such as the Delhi Laws (Special Provisions) Bill and the 2021 Master Plan. But they too have been forced into a corner by the visions of the forthcoming Commonwealth Games in 2010, and possibly the Asiad Games in 2014 and the Olympics in 2016. In addition, the rising political power of affluent (and aggressive) stakeholders, such as those organised into Manufacturers' and Traders' Associations, or Resident Welfare Associations and nongovernment organisations, under government-sponsored programmes such as "Bhagidari", have left little space for manoeuvre by governing institutions.

It is, therefore, within this context that one can situate the issues of transportation in the city.

TRANSPORTATION MODES

The official *Statistical Hand Book* of the Delhi government gives the following picture (Table 8.1) of the motorised transport in the city over the last three decades. The share between public and private transport has been estimated to be as given in Table 8.2. However, this picture clearly does not span the entire range of transportation modes available in the city, as is given in the following Table 8.3.

In other words, there is a significant presence of non-motorised transport within the city, which does not come to the attention of the transport planners, except for the purposes of licensing and regulating. It is also interesting to note that this summary of transport modes does not include walking and cycling either. This is all the more surprising since in the first Master Plan of 1962, there was a specific mention of cycles and provision was made for the construction of cycle paths. The mention of cycles disappeared in the second Master Plan (MPD-2001), although there was still provision for cycle paths. At the same time, another survey carried out by the Sajha Manch (with assistance from the Hazards Centre) in resettlement colonies,

Table 8.1 Growth in vehicles

Vehicle	1971	1982	1993	1996	1997
Cars and jeeps	61,521	134,084	510,242	685,850	705,923
Two wheelers	109,112	429,923	1,467,182	18,44,471	1,876,053
Auto rickshaws	10,812	23,396	71,568	80,208	80,210
Taxis	4,105	7,744	11,679	14,593	15,105
Buses	3,266	10,661	23,943	29,183	29,572
Goods vehicles	15,262	42,723	114,294	139,300	140,922

Source: *Delhi Statistical Hand Book 1988, Delhi Statistical Hand Book 1998.*

Table 8.2 Growth in daily passenger trips

Year	Total trips (lakhs)	Trips by mass transport (lakhs)	Trips by personal transport (lakhs)
1966	19.88	8.1	11.7
1981	39.0	23.4	15.6
2001	153.0	11.50	38.0

Source: Delhi is Doomed without Metro, Jag Pravesh Chandra.

Table 8.3 Man-/animal-driven vehicles in Delhi

Description	1980–81	1990–91	1992–93	1993–94	1994–95	1995–96	1996–97
Rickshaw	3,898	12,421	15,579	45,963	45,899	46,386	55,269
Tonga	1,822	974	927	867	796	679	613
Rehras	483	269	190	190	205	131	144
Hand carts	6,231	4,886	4,998	4,998	5,518	5,515	5,448
Bullock carts	695	521	442	442	423	426	430
Cycle trolley	3,815	11,476	24,637	35,576	38,925	42,339	40,666
Total	16,944	30,547	46,773	88,036	91,766	95,476	102,570

Source: Delhi Statistical Hand Book 1988, Delhi Statistical Hand Book 1998.

Table 8.4 Distribution of transport modes (%)

Transport mode	MPD-1962	MPD-2001	Sajha Manch
Private cars	8.0	16.6	n/a
Two-wheelers	2.0		2.0
Taxis/autos	n/a	3.6	n/a
Bus	30.0	59.7	31.0
Bicycle	60.0	17.3	39.0
Walking	n/a	n/a	22.0

Source: Master Plan for Delhi 1962, Master Plan for Delhi 2001, and Sajha Manch 1998.

Table 8.5 Dangers of work for lower-income groups (% respondents)

Hazard	Jhuggi Jhonpri clusters (slums)	Unauthorised colonies	Resettlement colonies, 1998	Resettlement colonies, 2006
Mechanical	2.01	33.33	20.99	11.20
Chemical	5.29	7.34	7.73	3.75
Thermal	0.74	6.21	7.18	13.40
Noise	0.95	12.43	10.49	10.10
Electrical	3.81	21.47	23.20	5.05
Other (travel)	87.18	19.21	30.38	11.05
Travel	–	–	–	49.75

slums, and unauthorised colonies in 1998 revealed that there was a significant number of people from these substandard settlements who were still walking and cycling to work (Table 8.4).

When asked about the dangers faced by them at work, a significant percentage, particularly from the slums, reported a category termed "other". On personal enquiry, several of the respondents said that this referred to the hazards encountered while travelling to and from work – a category that was not present in the questionnaire. A subsequent study steered through a United Nations Development Programme (UNDP) project by the Sajha Manch in 2006 in five resettlement colonies specifically included this category and, once again, it was observed that for this class of road users the fear of accidents and injuries on the road is very real and pressing (Table 8.5). But this concern finds no place in the lexicon of the decision makers and policy planners. Thus, it is evident that one entire class of stakeholders in transportation is missing in the perspective of the governing authorities.

PLANNED TRANSPORTATION

A quick glance at the Master Plans of Delhi illustrates the perspective of the planners with respect to transportation (Table 8.6). The cycle paths have actually disappeared from the latest Master Plan (MPD-2021), while

Table 8.6 Transportation provisions in the three Master Plans for Delhi

Planned norms for the Plan ending in	1981	2001	2021
Lakh trips per day	45	118	230
Vehicles	513,000	3,238,000	n/a
Buses	8,600	41483	n/a
Private modal share %	n/a	36	20
Public modal share %	n/a	60	80
Planned cycle paths	5	4	0

Source: Master Plans for Delhi for 1962, 2001, and 2021.

the shift of modal share to public transport is arbitrarily shown to be 80% in spite of strong evidence of a contrary trend. In fact, the Plan explicitly states that the "use of rickshaws has a direct relationship to migration" and, hence, cycle rickshaws are to be discouraged in order to prevent the undesirable migrants from entering the city. The Metro is regarded as being the factor that will bring about this miraculous change to public transport, although the data shows that the Metro caters to only 1.25 trips per day (tpd), as compared to the 23.4 tpd carried by buses.

As for the transport infrastructure, the MPD-2021 prescribes the construction of seven urban relief roads, several bridges on the river Yamuna, four interstate bus terminals, and five freight complexes by 2021. These are somewhat unreal prescriptions because there are no data and there is no analysis to show where and how these will be required and how they will cater to the needs of the city. For instance, a look at the three land-use maps of the three Master Plans (Figure 8.2) shows that in 1962 there were six crossings across the Yamuna, of which two were fords and pontoon bridges that were not usable in the rainy season. By 1982 (although the map was published only in 1990), the number of bridges was seven, with one carrying only railway traffic. In the 2002 map, the number of bridges increases to eight, and this remains the same for the zonal development plan for the riverbed, prepared in 1996.

These maps are particularly revealing because they indicate that in 1962 the riverbed area was coloured a light blue (coloured grey in the map) and assigned a land use as "flood plain". But in the second Master Plan, the colour becomes white or "not assigned". And in the third Master Plan other coloured sections begin appearing on the riverbed, including the proposed Commonwealth Games Village, indicating the future development of the entire flood plain for commercial purposes, as given in the Zonal plan. In other words, the underlying concept is that of commercialisation and sale of land for making high profits, and has nothing to do with either transportation or of catering to the needs of the citizens of Delhi. The same holds true for the "relief" roads, the bus terminals, and the freight complexes, because the Plan document provides no information on any studies having

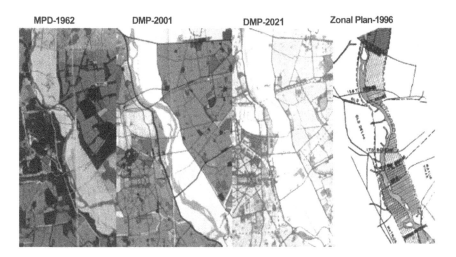

Figure 8.2 Bridges across the Yamuna and the changes in the flood plain.

Source: Master Plans of Delhi, 1962, 2001, 2021.

been done to identify where the greatest needs lie and how these needs may be fulfilled at optimum cost.

From other accounts of the planned development of the city, in fact, it appears that it is the forthcoming Commonwealth Games that is determining the transport requirements. Thus, of a total of Rs 770 crores earmarked for infrastructural development for the Games, as much as Rs 270 crores has been set aside for DTC to buy 1,100 dedicated low-floor shuttle buses, with automatic vehicle tracking systems, to link the airport, hotels, stadia, and tourist spots. An additional amount of Rs 265 crores has been set aside specifically for the PWD. DDA planners have said, "special care will have to be taken to ensure a smooth ride from the airport to the stadiums and Games village venues so that minimum time is spent on commuting". The contract for modernisation of the airport has already been awarded to GMR-Fraport for handling 80 million passengers by 2021. The railway stations are also going to be spruced up and their connectivity to the airport ensured to handle the additional visitors expected to come for the Games.

The ring roads are expected to become expressways, and a third ring road is proposed to provide access to the Games Village, as are two new bridges and a tunnel under the river. There will be several bypasses and underbridges as well as improvements to some of the key roads linking the different sites for the Games. There are plans for the construction of 24 new six-lane flyovers before 2009 to improve traffic circulation at a cost of about Rs 1,900 crores. These would be built to improve intracity connectivity, especially from the airport to the Games Village, the Village to the venues, the Village to hospitals, and so on. Apart from extending the Metro to NOIDA for the Games, there is also a proposal for the Metro to construct a high-speed corridor from New Delhi railway station to the airport for the Games. A high-capacity bus system would be started in seven corridors.

Clearly, all the above construction has been planned with an eye on commercial profits to be made from the host of athletes, managers, and tourists expected to come to Delhi for the Games. This is in spite of the cautionary warning issued by the International Commonwealth Games Committee that, "no country has ever shown a profit from the games" and that all host cities have been "warned that they were likely to incur a deficit". There have also been huge cost overruns in the past. Thus, Manchester in 2002, spent over four times the original bid and Melbourne in 2006 spent over five times its original bid. Melbourne had hosted the 1956 Olympic Games and finished paying off the debt incurred for that event only in March 2006, during the Commonwealth Games. So extensive has been the damage to the economy in other cities that the Manchester Council was prepared to give its Commonwealth Stadium away for nothing, while the Kuala Lumpur Commonwealth Games in 1998 provided a legacy of empty sports stadia, suppressed public demonstrations, and the policing of media coverage. But so strong has been the hold of some "stakeholders" over the government that all these warnings have been systematically ignored.

PUBLIC TRANSPORTATION

There is sufficient evidence to show that government agencies fall prey to pressure from lobbies in spite of evidence from expert bodies to take other more appropriate policy measures. Thus, on 28 July 1998, the Supreme Court of India passed a series of orders in a PIL filed by the lawyer, M. C. Mehta, on air pollution in Delhi, based on the expert recommendations of the Bhure Lal Committee (BLC), for full conversion of the entire bus fleet in Delhi to CNG by 31 March 2001. Subsequently, the date was extended to 2002. But, two years later, the data from the Central Pollution Control Board (CPCB) indicated that pollution levels for two parameters, nitrogen oxides and respirable particulate matter, had increased after 2002 (Figure 8.3).

Before one gets involved in the discussion on CNG being a cleaner fuel, one has to examine whether in fact the diesel buses were the real culprits. In 2001, only 6.7% of the total vehicles in Delhi were diesel-driven, the rest all ran on petrol. Buses constituted a mere 1.1%, although they carried over 60% of all motorised passengers. So it is not surprising that even if all the buses were converted to CNG, there would be hardly any dent in pollution levels. While these vehicles may be far more "polluting", their numbers do not add up to much for overall pollution levels. In addition, for each bus removed (with five round trips), 200 private petrol-driven vehicles would be required to carry the same number of commuters. In other words, the

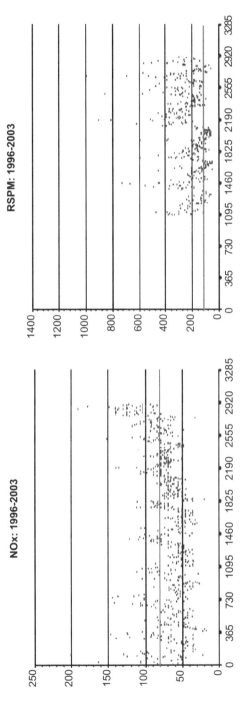

Figure 8.3 Nitrogen oxides and respirable particle concentrations after CNG.

Source: Central Pollution Control Board.

removal of 28,000 buses would entail a doubling of the number of private vehicle trips. Thus, the debate on diesel versus CNG appears to have been somewhat misplaced and designed to camouflage the massive impact of petrol.

Is all this in hindsight? Not surprisingly, the answer is no. For instance, just before the BLC came out with its recommendations, the papers of a World Bank Workshop on Vehicular Pollution Control were published. These papers not only analysed a range of options in fuels, lubricants, engine design, and technological upgrades, they also looked into traffic patterns, transport modes, enhancing public transport, and, most importantly, petrol engine emissions. Several of the authors also presented evidence before the BLC. Thus, the real question should be, "What were the cogent reasons that the BLC gave for rejecting the recommendations of all these experts?". Since the proceedings of the BLC have never been made public, nor has there been any transparency in its deliberations, we shall never know what these reasons were (even for future decisions).

But the lack of transparency may have had to do with the constitution of the BLC itself and the stakeholders it represented. It had as its members the Delhi Transport Secretary, an Environment Ministry official, a Petroleum Ministry official, the founder-Director of the Centre for Science and Environment (CSE), the CPCB Chairman, and, later, the Managing Director of Maruti Udyog (the car manufacturer). How many of these people were "technically competent" to decide on pollution is debatable. The CSE founder publicly wrote he had a "vested interest" in the issue because of his struggle with asthma and cancer – which he somehow related to "protecting the interests of the poor". Logically, then, other affected parties (such as the real poor, the commuters, and the employees) should also have been allowed space on the BLC, but they were not.

For instance, one of the clearly identifiable "stakeholders" could have been the bus drivers of the DTC itself enabling them to raise their concerns with diesel and CNG. A limited survey of 158 drivers by the Hazards Centre in 2006 gives pointers in this direction. The sample is a little skewed because 130 of the respondents were temporary drivers while only 28 were permanent employees. Nevertheless, the results give some clear indications of the manner of DTC's restructuring and its impact on the drivers. While the ostensible reasons given for restructuring are *efficiency*, *adequacy*, and *self-sufficiency*, the data on the drivers shows that the means employed to achieve these objectives were self-defeating – and were so for earlier attempts to reform DTC as well.

Table 8.7 illustrates that, while very few of the temporary drivers have work experience in DTC for more than three years, all the permanent drivers have been in their posts for over 20 years. The reason is that DTC stopped the recruitment of permanent staff 18 years ago as part of the process of restructuring and cutting down on wage costs. The implications of this on the wages of the drivers is clearly visible in Table 8.8, which shows

Table 8.7 Distribution of work experience among DTC drivers

Drivers (%)	Years of working in DTC								
	0–1	1–2	2–3	>3	20–22	23–25	26–28	29–30	No reply
Temporary	24.6	31.5	40.0	2.3	0.0	0.0	0.0	0.0	1.5
Permanent	0.0	0.0	0.0	0.0	39.3	32.1	14.3	14.3	0.0

Source: Hazards Centre, 2006.

Table 8.8 Monthly income of DTC drivers

Drivers (%)	Average monthly income (Rs'000)								
	2–3	3–4	4–5	5–6	10–11	11–12	12–13	13–14	>14	No reply
Temporary	42.3	46.1	6.1	0.7	0.0	0.0	0.0	0.0	0.0	4.6
Permanent	0.0	0.0	0.0	0.0	21.4	25.0	21.4	17.9	14.2	0.0

Source: Hazards Centre, 2006.

that the temporary drivers generally earn less than Rs 4,000 per month, while the permanent drivers have an average pay of Rs 12,000 per month. This clearly benefits DTC, but it also places undue pressure on the drivers who have been recruited on a temporary basis. This pressure has apparently been further compounded by the change to CNG, as indicated by the drivers.

As Table 8.9 indicates, even though the temporary drivers have been working for far fewer years than the permanent ones, they are already victims of musculoskeletal and neurological disorders. The symptoms of respiratory problems are far more marked in the permanent drivers who have been exposed to diesel fumes for over twenty years, but they have already become apparent in the temporary drivers. The drivers report that the new CNG engine is hotter, but not more powerful, than the old diesel one and consequently their working conditions have worsened. This provides an insight into the kind of pressure these drivers are operating under, and this will obviously affect DTC's performance as a whole.

These trends of contractual employment, privatisation, and inappropriate investment of public funds are also clearly evident within the other arena of public transport, that is, the Metro. A study of the Metro carried out by the Hazards Centre reveals that the environment impact assessment (EIA) carried out by RITES in 1995 is not applicable to the present corridors of the Metro, because they comprise only half of the planned sections for which the EIA was carried out, while the distance is greater (Table 8.10). The EIA report was never brought into the public domain and there has been no public participation in the plan. In fact, there has been no public or expert review to look into the several methodological flaws

Table 8.9 Illnesses reported by DTC drivers

Drivers (%)	Musculoskeletal	Dermal	Ophthalmic	Cardiovascular	Gastrointestinal	Respiratory	Neurological	Others
					Illnesses			
Temporary	45.3	0.0	3.8	2.3	7.7	6.9	26.9	25.3
Permanent	67.8	0.0	3.5	17.8	17.8	71.4	21.4	3.5

Source: Hazards Centre, 2006.

Table 8.10 Proposed and actual routes of Metro

No	Proposed (modified) Phase I Section	Km	Actual (constructed) Phase I Section	Km
1	Vishwavidyalaya – ISBT	4.5	Vishwavidyalaya – Secretariat	10.84
2	ISBT – Connaught Place	4.2		
3	Connaught Place – Secretariat	2.3		
4	Shahdara – ISBT	6.4	Shahdara – Tri Nagar – Rithala	22.06
5	ISBT – Shakur Basti	10.6		
6	Shakur Basti – Nangloi	8.0		
7	Subzi Mandi – Siraspur	12.8	Indraprastha – Barakhamba – Dwarka	32.10
8	Siraspur – Holambi Kalan	6.5		
	Total	55.3	Total	65.00

Source: EIA for Integrated multimodal mass rapid transport system for Delhi, http://www.dmrc.delhigov.in, 2006.

contained in the EIA. Thus, various important decisions were taken but not exposed to the public gaze.

Similarly, the majority of the amount invested in the Metro has been generated through loans. In the present scenario, the Metro is not in a position to cover its operation costs, let alone pay back its returns. It is currently running at a loss, and the trends show that in the future it will continue doing so (Table 8.11). Eventually other heads of government funds (public money) would be exploited for the repayment of the existing loans. On the other hand, the Metro is meeting only one-third of its revised expected ridership and one-fifth of its claimed ridership of 1995 because of the inequitable fare structure, no concession scheme, and the distances from the Metro station. This means that the Metro as a mode of public transport is only catering to the needs of a more affluent section of the public, which is ready to pay more for its travel.

Table 8.11 Profit and loss account of Metro

Item	Amount (Rs crores) 2001–02	2002–03	2003–04	2004–05
Income		5.9	46.6	72.2
Expenditure	67.6	5.8	32.0	52.2
Profit before depreciation and interest	1.7	0.1	14.6	20.0
Profit(+)/Loss(−)		−8.3	−32.4	−76.3

Source: (1) Annual Reports, DMRC, 2006; (2) Metro hurtles into financial abyss, Times of India, 22 May 2006.

Hence, in order to make up for its losses, the Metro has to go in for extensive property development for commercial purposes and the DDA has to declare a 500 m belt next to the Metro routes as a "development corridor" while permitting high-rise constructions. This further adds to congestion on the routes and defeats the very purpose for which the Metro was set up. In addition, the buses along the routes have to be diverted or curtailed or even cancelled (Table 8.12), which means they cannot compete for passengers with their lower fares, greater flexibility, and commuter convenience.

It is quite clear from these issues that the Metro is a part of a larger agenda driven by a select group of "stakeholders" to transform Delhi into a "world-class city" for facilitating and encouraging the investment of global capital into the city. The large-scale development of property on both sides of the Metro lines is an indicator that it has not really been brought into the city to provide better transport options for the commuter. Eventually, in the name of fast, efficient, and pollution free "public" transport, the Metro benefits only a small section of the "private". This transfer of public money into private pockets and distribution of social and environmental costs over a much larger population, who will not even travel by Metro, has been

Table 8.12 Bus routes affected because of Metro construction

	Route no.	No. of buses
DTC buses affected		
Discontinued	247	
	132	
	167	
Curtailed	817	13
	832	10
	405	10
	805	2
	61	1
Extended	778	3
	801	4
	915	2
	927	1
	968	1
Diverted	233	1
	917	1
Total	**15**	**58**
STA buses affected		
Curtailed	832	29
	817	61
Cancelled	247	16
Total	**3**	**106**

Source: Data acquired from DTC through RTI, 2006.

quite systematically camouflaged under a huge propaganda barrage by the government and the media, and reveals the true nature of "stakeholder" participation in governance.

OTHER ROAD USERS

It would also now be useful to look at some of the other "stakeholders" who are on the roads for the purpose of their livelihoods, and to what extent their concerns are incorporated into transportation planning. One of these road user groups is the three-wheeled scooter rickshaw (TSR). In a survey by the Hazards Centre in 2002, 57% of the respondents had bought new CNG-powered vehicles, 14% had had CNG kits retrofitted into their old engines, while 29% continued to use the petrol vehicles. It was revealed that most of the drivers of the TSRs are also the owners and that there are few who take the vehicle on rent (Table 8.13). Most of them drove between 100 km to 150 km in a day and their daily income averaged Rs 200. Thus, the TSR represents an important source of self-employment as well as a very convenient mode of paratransit for a large number of commuters within the city. They should, therefore, be encouraged through policy measures.

A TSR is preferable to a car as it carries the same number of people on an average, takes up one-third the parking area and one-half the space while travelling. Since it weighs one-third of a car it wears out the road much less, has less tyre/rubber use, and uses one-third of national resources to produce it. All this reduces indirect pollution. Since TSRs have a small engine they pollute much less per passenger than a car if the engine is as specified. Because of the small size of the engine, they can't go faster than 50 km/h, thus keeping to urban speed limits, controlling others' speeds, and reducing the number of fatal accidents among pedestrians and cyclists as compared to cars. However, the conversion to CNG because of stringent policy measures has not been without its adverse impact on the TSRs (Table 8.14).

Apart from the problems with the conversion to CNG, as reported by the TSR drivers, there are several other issues that affect them, and these are directly driven by the policies adopted by the government towards this class of road users. Some of these issues relate to the issuance of permits and clearances, the corruption prevalent in the Department of Transport, the low rates prescribed by the authorities, the absence of proper facilities

Table 8.13 Owner-driver characteristics of TSRs

No.	Response of respondents	% Owners	% Drivers
1	Yes	64.9	84.2
2	No	25.9	8.4
3	No response	9.1	7.3

Source: Hazards Centre, 2002.

Table 8.14 Problems with CNG

Problems	%	Problems	%
CNG kit retrofitted		New CNG vehicle	
Technical problem	61.5	Technical failure	37.9
Consumes more oil	5.1	Costly maintenance	36.5
Gas is not available	5.1	Gas not available	10.9
Lots of problems	5.1	Lots of problems	14.6

Source: Hazards Centre, 2002.

for parking and rest, the non-availability of repair workshops and skilled mechanics, harassment by the traffic police, the non-functioning of (tamper-proof) electronic meters, and the high costs of operation and maintenance (Table 8.15 gives the suggestions offered by the drivers to address the problems enumerated by them). What is important to note is that over 40% think that low fares are at the root of the conflict between customer and TSR driver, followed by 36% who are concerned about corruption in the Transport Department. And as many as 10% are seeking to be heard as bona fide stakeholders by government.

Another class of vulnerable stakeholders is that of the cycle rickshaw pullers. A series of interviews with 50 of them by the Hazards Centre in 2006

Table 8.15 Policy suggestions by TSR drivers

No	Suggestions	%
1	Increase rates to prevent conflict between driver and customer	41.1
2	Zonal offices; no agents; compulsory identity cards; pass vehicle after completion of papers; reduce fees to curb harassment by Transport Department	35.9
3	Provide proper facilities at increased number of stands to increase utilisation	23.4
4	Issue tamper-proof certificates for meters to fix regulator's responsibility	15.8
5	Reduce cost of vehicle/spare parts; provide institutional finance for employment	11.0
6	Mechanics and increased gas supplies to improve turnover and efficiency	10.5
7	Access to government to promote discipline, single union, and public dialogue	10.0
8	Replace meter and permit system to reduce red tape	9.1
9	Avoid CNG kit as it has too many associated technical problems	6.7
10	No penalties for parking and adequate parking areas for convenience of passengers	5.2
11	Close prepaid counters and promote payment by meter to curb touts and corruption	3.3

Source: Hazards Centre, 2002.

discovered that three-quarters were in the prime working age range of 20 to 40 years, while over half were illiterate. But, unlike the TSR drivers, the vast majority (over 90%) took the rickshaw on rent, typically Rs 25–30 per day, travelled on random routes according to the requirements of the customer, ferried an average of two passengers per trip, most of whom were families, and travelled over 30 km in a day. Two-thirds earned more than Rs 2,000 per month, which was reasonably more than the rent they had to pay the owner, and were not in favour of restrictive licensing of cycle rickshaws.

More than half felt they had to park wherever they could find space, as there was no demarcated parking, and complained of routine harassment by the police and the municipal authorities on this account. These authorities would either puncture the tube or confiscate the rickshaw itself, thus making earning a livelihood all the more difficult. Plagued as they are by the charge that they are responsible for congestion on the roads, 90% favoured a separate lane, but more than half argued that traffic jams occurred primarily because of wrong parking by cars. Yet this class of road users, like that of the TSR drivers, has no voice in governance nor do they have the opportunity to present their case when transport policy is being formulated.

Finally, we present the case of the waste pickers, who form an important link in the informal chain of recovery and recycling that is part of the economy of the city. Not only do these waste pickers forage on the sides of the roads, and occasionally live there too, but the entire transportation of waste is a matter of grave concern for them because it also constitutes a part of their overall illegality in the eyes of the ruling establishment. Not only does the waste legally belong to the municipality (and, therefore, they cannot officially pick it up), but their source of livelihood also gives them an appearance that makes them easily prosecutable under the Beggary Act or the Foreigners Act.

The waste picker not only forages in the markets and at the collection points or open sites for material that has a value in recycling, they also have to sort the material into different categories before selling to the kabari or junk dealer. In a study conducted by Chintan, with the assistance of the Hazards Centre, in 2002–03, 54.4% of the respondents who were making most of their collection in the NDMC area said they were segregating their waste in or in front of the kabari's godown, while 33.1% conducted this activity on the footpath. But closeness to the eventual buyer was obviously critical for the trade itself. The waste pickers were collecting an average of 69 kg of waste per day. The mode of transport for collecting the waste was mostly cycles, followed by walking, and cycle rickshaw (Table 8.16). This is understandable because NDMC authorities do not permit cycle rickshaws within their area.

In another phase of the study, in the MCD areas, where rickshaws are permitted on payment of Rs 360 per year, it was discovered that the waste pickers covered a much larger range of ground. Only 152 of their trips were into the NDMC area, while 826 trips were made within MCD territory. Because of their direct association with kabaris, a much higher percentage (90.5%) were segregating their waste inside, near, or outside the godown.

Table 8.16 Mode of transport for waste pickers, 2006

Waste pickers (%)	Area		Thiawalas (%)
	NDMC	MCD	
Walking	24	29	69
Cycles	58	19	31
Rickshaws	18	51	–

Also, there was much greater use of rickshaws than cycles because of the relaxation of permits in the MCD area (Table 8.16). Of the sample, 90% were single wage earners and, on average, they were collecting 57 kg of waste daily, which is quite comparable to the NDMC sample. Earnings of this group averaged Rs 90 per day, which is significantly lower than the minimum wage for unskilled labour (Rs 127) stipulated for Delhi.

Thiawalas are intermediary collectors who operate out of a "thia" in markets and other central locations. They do not go to collect the waste but the waste generator comes to them for sale of the waste. Thus, they do not need to use cycle rickshaws but prefer to walk as their personal mode of transport (Table 8.16). In the context of the waste pickers, though, what is of great significance is the correlation between the mode of transport, the loads that can then be transported, and the related earnings. As the survey data shows, the waste pickers who operate on foot are largely able to carry less than 40 kg of waste on one trip, have a range of 6–7 km, and earn Rs 50 daily. Those who have cycles are mostly transporting between 40 to 60 kg over 20–25 km and earn Rs 100 per day. The rickshaw operators load 40–100 kg in one trip, but travel 10–15 km.

Since the distances and territory that the rickshaw operators cover are also determined by the restrictions placed on rickshaw movement by the municipal and police authorities, this becomes an important issue of transport policy. The ability to enhance earnings is also, therefore, dependent on the mode of transport that the waste picker is able to use. Quite clearly, this is related to the extent to which the waste picker is able to get formal recognition and space in the design of civic life. However, the waste picker (and the associated kabari) is considered to be on the lowest rungs of the social ladder because of the vocation they pursue, and legitimacy is not granted by the authorities. On the contrary, judicial orders spurred by mischievous "public interest" litigation has seen to it that the waste picker is further criminalised and marginalised.

CONCLUSION

The evidence presented in this chapter shows that the procedures of governance in the city are conditioned by the variety of elected, nominated, and bureaucratic institutions that contend for supremacy in decision making. In the current period it is clearly the judiciary that has taken the lead in steering both policy as well as implementation, mainly based on a debatable interpretation of what constitutes "public interest".

Within such a context, there has been an increased focus by policy makers on private motorised modes of transportation that exclude large sections of the people who are dependent on personal non-motorised or public transport vehicles. In particular, walking and cycling, which are the most preferred modes for the poorer sections of the population, are almost completely ignored in transport planning.

Such exclusion is evident in the manner in which the Master Plans of the city have been formulated over four decades and how the cycle path has disappeared while the pedestrian finds little or no mention. In addition, mega events, such as the Commonwealth Games, have taken over the imagination of the city and all transport planning seems to be directed at how to transport the athlete and the tourist across selected parts of the city as rapidly as possible.

In the arena of public transport, policy has been driven by issues of pollution and congestion. Hence, the conversion of the DTC fleet to CNG and the construction of the Metro routes have actually ignored the needs of the ordinary commuters who use public transport. Consequently, both have had adverse impacts on the life of the working population and will, in the long run, prove to be counterproductive for the economy of the city.

Other vulnerable road users, such as the auto rickshaws, the cycle rickshaws, and the waste pickers, have also been eased out of the perspective of the planners, although they contribute in significant ways to the mobility and health of the city. Hence, there appears to be a deepening gap between the institutions of government and the people. Policy is catering to the needs of a select few within the population who are wealthy enough to be able to meet the increased costs as well as powerful enough to influence government.

In conclusion, it is clear that one set of "stakeholders", among the many, is calling the shots, and is railroading the rules to strengthen its convenience, its profits, and its control.

Chapter 9

City governance and effectiveness

Gerald Frug

CONTENTS

INTRODUCTION

I would like to talk today about how to think about organizing governance mechanisms that deal with urban transport. The problem facing transportation planners, I suggest, is figuring out how to combine the general and the particular, the broad and the narrow. This problem comes up in three different ways. Consider first the definition of the subject matter for planning. The broad way of understanding the topic is to recognize that one cannot talk about transportation without talking about other urban problems at the same time – above all, land-use planning, affordable housing, business location, and sprawl. Even if one focuses solely on transportation, the broad way of talking about transportation is to focus not just on roads and mass transit but on taxis, walking, bicycling, cycle rickshaws, jitneys, and many other ingredients of the system. The second aspect of the problem is the breadth of geography we should be focusing on when we plan: is it the neighborhood, the ward, the city, the region, the state, the nation as a whole, or even, as with air travel, larger still? How broad, and how narrow, should we be thinking? The final topic is the amount of participation in decision making about the subject matter (however defined) for the geographic area (however defined). Whose voice is entitled to be heard? What role does the public-at-large – democracy – play in thinking about transportation? What role do experts play? What role do stakeholders play? And, if stakeholders play a role, who, exactly, are the relevant stakeholders?

The problem presented by these three questions is not that we don't know the right answers to them. At the conceptual level, we do know the right answers to all three issues. The topic should be inclusive, not just focused on transportation. Of course, a specific focus on each element of transportation is necessary too. But these specific analyses have to be connected to each other and, then, to urban planning more generally. The geographic scope similarly needs to be both broad and narrow. We have to be able to think beyond the neighborhood and the ward to the city, the region, the state, and the nation – and, at the same time, understand how to organize transportation at the level of the ward, the neighborhood, and individual streets. Finally, transportation decisions will be illegitimate unless they respond to local residents' needs and desires – respond, in other words, to the democratic will. But this does not mean that there isn't a role for experts too. Of course, experts have a role. But expert decisions have to be connected to democratic decisions. In thinking about all three issues, in short, we need to be broad and narrow at the same time.

This conventional wisdom is easy to state. But, as far as I know, it is not implemented anywhere in the world. Virtually everywhere, transportation decisions are fragmented into bits – both functionally, and as a matter of geography. Transportation decision-making is rarely connected to other aspects of urban planning. In fact, different aspects of transportation are run by different agencies – some of a city's roads are subject to the juris-dictions of the city itself and others to state or national control; buses are operated by one agency and commuter trains by another; taxis are regu-lated by a different government agency altogether. Pedestrians and bicycles are usually left out of the picture altogether. Geographic fragmentation is equally intense. Different parts of the transportation system are allocated to the national government, to the state government, to the city, to many different specialized public corporations and agencies, and to the private sector. No one puts all these parts together. Finally, popular participation – if it exists at all – plays only an intermittent and occasional role. The two papers presented in this session demonstrate that this fragmentation exists in India. Consider simply the title: Is Anyone in Charge? I found the two papers very interesting and thought-provoking. I must admit, however, that I found their descriptions of who is making decisions about transportation in India too complicated to follow.

There is nothing unique about the situation here in India. Consider simply the organization of transportation in New York City. The two most impor-tant actors on transportation issues are the Metropolitan Transportation Authority and the Port Authority of New York and New Jersey. The Metropolitan Transportation Authority is appointed by New York State's governor, with only 4 of its 17 members recommended by the city; the Port Authority is appointed by two governors, without any city input. The Metropolitan Transportation Authority runs New York City's subways and buses, along with the Long Island Railroad; the Port Authority runs the airports, PATH trains to New Jersey, and the Air Train at John F. Kennedy Airport (JFK); New Jersey Transit, appointed by New Jersey's gover-nor, runs its own trains and buses into New York. The Transportation Authority operates nine bridges and tunnels; the Port Authority controls other bridges and tunnels, including the Lincoln Tunnel and the George Washington Bridge; the New York City's Department of Transportation controls still other bridges and tunnels, such as the 59th Street Bridge. The highways are run by the New York and New Jersey State Departments of Transportation. And the New York City Taxi and Limousine Commission licenses the city's taxis. Transportation, you should know, is the area for which the federal government in the United States is most insistent on metropolitan planning. The problem for New York is that there are many metropolitan transportation planning bodies in the area, not just one. One deals with New York City and a few nearby New York suburbs; another deals with New Jersey; yet another deals with Connecticut. No one, starting from scratch, would devise a transport and mobility structure like this one. To declare this set-up a scandal would be a waste of time. Everyone knows it's a scandal; it's been a scandal for decades.

Why is transportation policy organized in this fragmented way? Why can't we design a system that is not a scandal? The crisis facing the organization of transportation is not deciding what would be better – we know what would be better. The crisis is figuring out how to get from here to there. We don't seem to know how to design the broad-based system we think desirable. And we don't seem to know how to relate such a system, even if we had one, to the equally important narrower focus. We need, in other words, to figure out how to relate parts and wholes. Even more urgently, we need to figure out how to begin to head in the right direction.

I certainly cannot answer this question for Delhi. I cannot answer it for any city. But here are a few initial thoughts. The crisis in transportation planning is connected to the crisis of thinking about cities generally. The fracturing I've been emphasizing on the issue of transportation policy occurs on every issue. To begin to address this problem, we need to organize three institutions. First, we need to organize neighborhood-based thinking about transportation issues and about how these issues relate to others. At the same time, we need to organize city-wide and region-wide thinking about the same topics. This will require a lot of hard work at every geographic level. But we cannot even begin to discuss the matter until people think about what to do. Even harder than this thinking process, however, will be the effort to relate these conversations to one another. There will be no consensus within any particular neighborhood, no consensus joined by all the neighborhoods, and no consensus between the city and the region. There therefore has to be an ultimate decision maker – some body that can work out the necessary compromises. One decision maker has to have authority over the countless agencies and departments now involved in these issues. This ultimate decision maker should represent the largest relevant area – the region or the state. But, to be legitimate, the ultimate decision maker has to be organized in a way that makes it representative of the collective voices of its constituent cities and neighborhoods. Like a legislature, it has to be organized in a way that allows it to work out compromises between different parts of the city and region, not simply come up with its own decision based on experts or stakeholders. This means that the experts have to work at every level, not just at the top, and, at the same time, that democratic voices have to be heard at every level, not just at the bottom. The locals need to be taught by the experts; the experts need to be taught by the locals. Regional thinking needs to be brought down to the more local level, and all of the local levels, collectively, need to be constituent parts of the regional level. This is an organizational issue – an issue of institution-building.

Implementation of such a scheme cannot happen overnight. Indeed, it has to consider more than simply transportation. But we can't wait to set the whole process up before we begin to do something about the transportation problems we face. Perhaps it's best, then, to begin with a single issue and try to make some progress on it. We need to organize institutions that deal at the neighborhood, city, and regional level with this one issue – and, once they make some headway on it, that can expand their jurisdiction to other issues. I have no idea what the issue here in India should be. That needs to be discussed by people here. But I think the issue should be concrete – something at the level of reducing overcrowding on trains. The task is to begin the process of relating the parts and the whole. Without that, we will have no legitimate decisions about transportation. In fact, we might have no decisions at all.

Chapter 10

Alternative transport policies for personal public transport

Lessons learned

Hermann Knoflacher

CONTENTS

INTRODUCTION

There is no one in any city who does not say that provision of public transport is the most important aspect of transportation planning. But the large investments and up-promoting [of] personal transport do not provide choices for a majority of citizen[s] due to the overwhelming dominance of private cars. This was the case also in most of the European cities until the 1960s or 1970s of the last century. Practitioners have learned to handle the system in a better way by trial and error in many cases.

The main problem is that traditional transport planning is not dealing with people, it is dealing with cars, vehicles, carriageways, and rails. This contradiction between expectations and reality indicates that the behavior of the system has not been understood. There was no distinction between system behavior and human behavior and no knowledge that the system behavior might be different from human behavior. Evolutionary epistemology was not the method used in transport science and urban planning. To understand human behavior in this artificial environment it is necessary to understand human evolutionary structure and conditions. Empirical and theoretical research have been carried out over the last 40 years at the

Institute of Transport Planning and Traffic Engineering at the Technical University in Vienna, providing tools for a better understanding of system and people's behavior. Based on these empirical and theoretical findings new solutions have been found[1]. These findings applied in practice show how effective this new kind of treatment of the sickness of the cities is. This paradigm change is necessary to prevent mis-investments and to guarantee choices for the majority of citizen.

If we want to realize alternative transport policy for personal public transport it is not enough to look for so-called "best practice" solutions, because most of today's best practice solutions are "better practice of the worst cases". We have to use the best scientific knowledge. Unfortunately a paradigm change is difficult, but necessary.

There are several reasons for this phenomenon:

- Worldview of transport engineers and transport experts: For the last 50 years approximately, transport engineers have been trained to optimize the transport system for car users, following the American "bible for traffic engineering", the *Highway Capacity Manual*[2]. Engineers are trained to build physical structures. It is much easier to build a physical structure for car traffic than for public transport.
- Worldview of economists: They are trained to optimize "individual benefits" in a very narrow view. Everything that is outside their view is called "external costs".
- Worldview of architects as urban planners: Under the influence of the Swiss–French architect Le Corbusier (1887–1965), they are trained to build houses as single elements separated from everything else.
- Worldview of urban planners: They have followed the architecture-based *Charter of Athens*[3] and have separated urban functions without taking into account the effects.
- Politicians: They believe what these experts have told them and are often influenced by lobbying[4].
- Investors: They are looking for their short-term benefits, since nobody was asking them about the "external costs" of their activities.

These have been the ingredients of planning and decision making – more or less everywhere in the world – during the last hundred years, with increasing intensity in the second half of the twentieth century.

The only mode that can cover the large gaps between disciplines and separated urban functions and that has at the same time benefits for individuals was and is the car. But this transport mode produces extremely big disadvantages for all the other transport system users, the local economy – and the cities themselves. Since car drivers belong to the "upper class" of society, and all those belonging to the professions mentioned above are car drivers, they provide the infrastructure for themselves. The rest of the transport system is forgotten as soon as anybody is using the car. The non-motorized part of the population is supporting this kind of political decision, since they are following the American dream that everybody who has no car might or should have a car in the future. Politicians and experts promise that infrastructure for car traffic will improve their quality of life or at least improve the "economy".

Under these circumstances it is obvious that this scattered field of interest is an ideal background for lobby activities to change the minds of people about their benefits. Car manufacturers, the oil industry, the banking sector (including the World Bank), and the construction industry, with all their consultants, are beneficiaries from the road infrastructure. Road infrastructure is an infrastructure for an open and uncontrolled transport system. This is the basic difference between roads for cars and public transport. If somebody has access to an uncontrolled infrastructure they cannot be controlled by society. And this is exactly what investors want – to choose the cheapest opportunity (piece of land) to build their "camps", e.g. shopping centers, and have access via the road network to all parts of the city – in a totally uncontrolled way.

Road transport planning is much easier than public transport planning. Engineers and planners who have not learned to plan and manage the *transport system as a whole* prefer to do what they can easily do – planning and building roads, carriageways, expressways, and motorways. If they have to plan public transport, they do it in a similar way. They make projects for public transport lines and stations, without taking into account the whole transport system and the effects from and on urban structure and economy. Public transport has to be planned as an integrated part of the whole urban economic and social system.

Decision makers are in general car drivers and not users of the public transport. People in the media are car drivers and have car traffic on their minds and not public transport or people's needs.

And finally there is a big difference between words and actions. Talking about public transport is popular and easy, but realizing and using public transport is a challenge for everybody. Only highly qualified experts, engineers, and politicians are therefore in a position to build not only public transport, physical infrastructures or railway lines, but also to understand the transport system and set the right measures and actions in place for a successful operation of the whole. Public transport planning is not a traffic engineering task alone, it is not an urban task alone, it is a challenge for both and more.

CITIES AND TRANSPORT

1. Is there a limit to motorization in the "ideal" city?

There are several answers to this question. First of all there is nothing like a so-called "ideal" city. We can only talk about the "normal city". The normal city is an entity that provides society with a built environment that contain everything the people need, with the lowest amount of physical transport needs. The basic needs of people are safety, health, subsistence, security, social integration, and employment. These are the ingredients of a normal city that have to be connected in the most efficient and effective way. This also has to be done in the public space. Public space is precious and has, therefore, the need for multifunctional use. The city center is a vital and vibrant part of the urban body, which is connected by public transport to other, similar, attractive urban centers, if the city is big. The degree of motorization of a sustainable city can be as low as about 50 cars per 1,000 inhabitants, or even less if the city is well organized. An empirical analysis of urban road networks and motorization provides a figure of about 0.5 m road length per person for a normal city, compared to 8 m/p in US and Australian cities[5].

Mankind has had experience of normal cities for about ten thousand years. These long-living, urban structures are the "normal" city. The normal city has a pedestrian public space and an environment with a high density and a variety of opportunities for human activities on the ground level of buildings open to the streets[6].

The increase in motorization damages and destroys the normality of a city. Public spaces is lose their multifunctionality and are cut into by carriageways for high-speed movement, parking, and sidewalks. The life-supporting network of the urban body, which has provided the "connectivity" of the city, is destroyed for the benefits of high-speed and long-distance transport. Cars and car users create an abnormal city: with an anonymous urban sprawl and a "city center", about which American urbanists say "The city center is there where not there is there"[7]. The degree of motorization is an indication of the structural and logistic deficits of a city. What planners have forgotten has to be compensated for by car mobility of the population and make the city dependent on a continuous inflow of gasoline.

What about big cities?

The "normal" city can grow to about one million inhabitants if the environment is fertile enough for all the supplies the city needs.

With an increasing number of inhabitants society has two options:

- to develop a so-called satellite city (a Greek strategy for hundreds of years)[8] or to introduce an artificial transport system.
- An artificial transport system can use either energy-efficient mobility, such as bicycles, or rely on external energy, such as all modes of public transport and/or cars. Cycling can extend the city size to up to 10 million inhabitants without creating any problems for sustainability. With the introduction of public transport, rail-based or not, further growth of the city is possible with the help of artificial energy

(electricity, coal, gas, and oil). Therefore motorization can be close to zero, if cars are necessary only for emergency purposes or handicapped people and – to a certain extent – for the transport of goods. In a "normal" city the limit to motorization will be around this figure.

European cities can exist without any serious problems or decrease of quality of life with a degree of motorization of between 50 to 200 cars per 1,000 inhabitants (instead of 300 or more) if they have protected and modernized their historical urban structures and reorganized the new one in the proper way.

2. Can there be alternate models for personal transport, especially the use of motorcycles, which are very convenient and low cost, but very hazardous?

If the urban structure is "healthy" there is not very much need for mechanical transport. Motorcycles are not only dangerous, they pollute the air and are noisy – and private motorcycles are a part of the uncontrolled transport modes in the same way as private cars. Only small motorcycles for low speeds are possible, to a limited extent. If motorcycles were to pay the whole external costs they put on the society, it would not be a low-cost mode. Public electric powered motor rickshaws and cycle rickshaws are efficient and socially acceptable for the urban transport system

3. Under what conditions will private vehicle owners use public transport in the modern city with multiple business districts?

It needs to be more attractive to use public transport than to travel by car or motorcycle. The key measure is the organization of parking. This has not been understood in traditional transport and urban science. Parking organization at the origin and destination is the most effective measure to influence human behavior in the transport and urban system. (See Annex 1.)

Private vehicle owners will use public transport if the city introduces parking regulations that take into account real human behavior and that guarantee the possibility of choice for car owners. Parking at home shouldn't be allowed as well as parking at the destination. Parking should only be allowed in parking garages and nowhere else. The same system has to be true at the destination too[9]. Under these circumstances private vehicle owners have no reason to use the car for their normal daily trips, since they have no better access than with the public transport. Public transport will become the backbone of the transport system for longer distances between the different districts. Most of the roads can then be pedestrianized and made easy to use for cyclists. Under these circumstances the multiple business districts will very soon change to become closer to a "normal" city. Due to the great variety for people and business, the city will develop into a sustainable direction. Mental mobility, as an unlimited resource, will be the driving force for the future, instead of physical mobility.

4. What criteria do we need to use for determining choice between surface and grade-separated public transport systems?

Beside the "trivial" engineering indicators such as cost or space, the most important criterion is the acceptance of or accessibility for the user. Because the ultimate purpose of public transport is service for people. People's behavior is dependent on the energy available for movement[10]. Public transport stops that are separated by great distances and/or are only accessible by stairways, elevators and escalators reduce the catchment area of any one stop substantially compared to a surface public transport situation. Surface public transport is therefore the "normal" public transport to be used in a "normal city" for people.

One of the secrets for the success of good public urban transport is the visibility of public transport vehicles in the public space. The most important transport system is in the minds of people. If public transport is not visible and present in a public space it will not be present in the minds of people. Therefore cities with a visible public transport have a much higher share of public transport trips, compared to cities with a "hidden" and/or expensive underground system.

Transport policy is also to do with clear signals, the "flag" of "flagship". Only a public transport system above ground can fulfill this function – as well as being much cheaper and more efficient. Grade-separated public transport in places where it also can operate on the road surface indicate that the city has lost the battle against the problem of car traffic. The answer to this question can therefore not be given with technical and economic figures alone.

REFLECTION ON THESE QUESTIONS

These questions are sub-questions on a wider scale. To give the right answers we have to go one step up: *What should a transport system look like for a sustainable, economically viable and socially agreeable city?*

The transport system has a service function and is a means rather than the purpose. The purpose of the transport system is to serve the urban and human purposes: safety, security, economy, health, and beauty of a city as an attraction for people to live and work there and for companies to have their businesses there. The city has to be attractive not only for people but also for companies, investors, and tourists in accordance with its social structure and its long-term survival strategy.

The transport system is not a means to exploit the majority of people for the benefits of a few. It is not the means to make people ill from noise and air pollution, and it is not a means to disintegrate the urban structure – the visible, and invisible, urban body.

There is a flipside to the shiny surface of individual benefits, the individual enhancement of freedom of car users – a dark and dangerous flipside: it is the loss of control of urban and transport development.

A city is falling apart if the freedom of the individuals, whether people or companies, exceeds the freedom of the regulator of a city, the political or administrative body.

Currently, public transport problems cannot be solved within the public transport system itself, because cars dominate the city and the transport system due to existing regulations.

It is too late to try to regulate traffic flows, when parking regulation has not taken place in the right manner. If parking is individualized – the city is individualized and cannot be controlled any more.

To balance the degree of freedom between public transport and car traffic is only possible if parked cars are not more accessible than public transport stops. A city can only regain its regulatory function, if it can control and monitor its transport system in such a way that the number of parking opportunities (garages) in a city is much lower than the number of public transport stops in the city. And if no parking is allowed elsewhere in the city. The only exceptions are the so-called "working traffic" for loading and unloading goods and the needs of physically handicapped people.

Cars, the means of individual freedom, have to be separated from human activities, at least as far away as public transport stops are situated. A sustainable city must provide a greater choice within its system for people than for cars. Only if people have a greater choice within a system, will they have the opportunity to reduce the need of long trips. Traditional transport planning and urban planning is doing exactly the opposite.

For rail-based public transport, it is necessary to enhance the variety in its environment through regulation since it is a socially acceptable system. The same is true for other kinds of public transport modes. But individual cars or motorcycles have been restricted; otherwise they become the regulator, which is the case in most of the cities of today. If individual motor vehicles become the regulator they destroy the economic base of public transport, they damage the urban structure, and they destroy the social connectivity between people – and finally they destroy the security and safety of the urban space. If a city doesn't follow these basic principles, which are known as the law of requisite variety (or Ashby's law of requisite variety) the politicians and administrators will not be able to act as a serious regulator any more, they can only comment and try to catch up with the uncontrolled car traffic development and its followers or its effects.

There is a simple question to be answered by politicians, administrators, and experts: Do they give priority for the future of man and the city or do they give priority for the free movement of individually driven machines?

WHY SHOULD PUBLIC TRANSPORT IN URBAN AREAS OPERATE ON THE STREET LEVEL?

There are many reasons why public transport in urban areas is much more successful than elevated or underground solutions.

1) *Economic reasons*
    Public transport on the ground in form of streetcars (trams), rail-based or buses, is much cheaper to build, maintain, and operate.
2) *Efficiency*
    Public transport is one of the most efficient artificial technical modes of energy consumption, use of space, and safety. Therefore there is no reason to remove it from the road surface. The indicators of good, organized public transport systems concerning the use of space are similar to those of pedestrians and cyclists. They are 10 to 50 times better than car traffic.
3) *Accessibility*
    The efficiency of public transport is dependent on its accessibility. Accessibility is dependent on the acceptance of walking distances to and from public transport stops. In the vertical, 1 m is equivalent to about 15 m in the horizontal. If public transport customers have to overcome 6 m in height, the catchment area of a public transport system is reduced by at least 100 m in its radius. Elevated or underground public transport is therefore losing half or even two-thirds of potential customers compared to street-level public transport modes. If public transport is separated from the street level, it is necessary to build and operate escalators, lifts, etc., which is the common standard of modern grade-separated public transport modes. This increases the costs of construction, maintenance, and operation.
4) *Social issues*
    Public transport on street levels needs to be easily accessible for elderly and handicapped people, for children, and for those carrying goods.
5) *Security*
    The whole transport system on street level is under public social control. Public transport on the street level is therefore much safer. There is an interrelationship between public transport system users who have the view on the urban environment from one side and, from the other side, there is the view from shop owners, residents, pedestrians, and cyclists.
6) *Urban economy*
    Street-level public transport is good for the urban economy. The experiences in European cities show what is happening: where street-level public transport has been replaced by underground systems, it has had negative effects on local shops. Underground systems, like metros, change the economic structure of a city. Shopping streets are dying out between the stops of metros and shops are concentrated around metro

stops. Streetcars or buses are in total accordance with the need for local shops and urban life, and they keep the balance between mechanical transport modes and people's needs. Underground or grade-separated public transport systems increase the disparities and therefore the need for longer travel. They work in favor of big international chains, but damage the variety and richness of local shop owners.

7) *Structural reasons*

Public transport at street level keeps people moving without fundamental changes to urban structures being needed. Large, separated infrastructures in public transport have a strong influence on structural changes. The winners are the landowners around public transport stops, and the losers are the rest of the city.

8) *Urban reasons*

The visibility of public transport on the streets is an excellent "picture of a business card of a future oriented, environmentally friendly human-scale city". If public transport doesn't appear on postcards, it also doesn't appear in the minds of people. It is an important urban issue to integrate public transport into the mental map of people and visitors. Public transport on the streets tells the people that it is a socially balanced and agreeable mode of travel, available for everybody, easily accessible, and integrated into the urban body.

9) *Environmental reasons*

Finally there are environmental reasons for public transport to be operated using environmentally friendly kinds of energy. This is the case in most of the European cities today. Public transport on the street level is a kind of indicator for the environmentally friendly transport policy of the city. "Environmental" doesn't only mean nature, it also means the human environment that has to be integrated with, and not separated from, public transport. To integrate public transport in human society it is necessary to keep it on the road surface instead of putting it into the sky or under the ground.

The only reason to push public transport from the street level:

There is only one indicator that is against public transport at the street level: "priority for cars instead of priority for people", joyful car driving instead of an efficient urban transport system.

THE TRANSPORT SYSTEM HAS A SERVICE FUNCTION AND IS A MEANS, NOT A PURPOSE

Public transport problems cannot be solved within the public transport system since the cause of the problem is the wrong physical structure of the car system, including the way parking is organized. To give people an opportunity to choose between cars and public transport, it is necessary to park cars in garages or to have parking places at least as far away from human activities as public transport stops are situated.

- The number of garages for cars (parking opportunities) in a city should not exceed half the number of public transport stops in the city in the transition period. The final solution is car-free cities.
- Public urban spaces can only be used freely by pedestrians and cyclists.
- For the transformation of the problematic system of today it is necessary to introduce financial regulations in such a way that parking at home has to pay for all the costs it produces. Parking at the right place has to be at a lower cost.
- The parking place of a car has the same effect on space and time as a public transport stop. Therefore parking cannot be treated as an individual right. Parking has to be controlled by society and can be only situated in places in accordance with the accessibility of public transport.

REFERENCES

1. Knoflacher H. 2007. *Grundlagen der Verkehrs- und Siedlungsplanung*, Böhlau Verlag, Band 1 Verkehrsplanung, Wien.
2. *Highway Capacity Manual.* 2010. Transportation Research Board and Academy of Science, 1950, 1965, 1985, 2000, 2010. Transportation Research Board (TRB).
3. https://en.wikipedia.org/wiki/Athens_Charter
4. https://ec.europa.eu/transport/themes/strategies/2001_white_paper_en
5. https://www.uitp.org/sites/default/files/Full%20dataset%20description%20-%20web_0.pdf
6. Knoflacher, H. 1996. *Zur Harmonie von Stadt und Verkehr. Freiheit vom Zwang zum Autofahren. 2., verbesserte und erweiterte Auflage.* Böhlau Verlag, Wien – Köln – Weimar.
7. Comment of an American urban planner in a TRB meeting in the 1990s in Washington.
8. https://www.ancient.eu/Greek_Colonization/
9. Knoflacher, H. 1981. Zur Frage des Modal Split. *Straßenverkehrstechnik*, (25)5: 150–154.
10. Knoflacher H. 1980, Mai. *Öffentliche Verkehrsmittel - neue Strukturen zur Verbesserung ihrer Chancengleichheit im städtischen Bereich.* Internationales Verkehrswesen.

ANNEX I

Austrian Society for Transport Science
Instruction for Practical Use
Group for Intermodality

Institute for Transport Planning and
Traffic Engineering
University of Technology; Vienna

Parking planning instructions for fair conditions between transport system users program for sustainable cities

Agreed version 2005

1. Preliminary remarks

Equal treatment of transport requires the knowledge of mechanisms of action in traffic systems. The technical traffic system enables the access to high and effortless speeds. This applies for public transport as well as for cars. With increasing motorisation, the share of public transport in passenger transportation decreased (Germany 1950: 65% public transport, today 17%). For equal opportunities of technical carriers, not only the number of parking spaces, thus motorisation is important, but also their site in relation to comparable elements of public transport.

The parking space is equivalent in its function to the stop of public transport.

The mechanisms of action on the behaviour of human beings, which are carried out by the car are so drastic that only a physical change of structure can bring about equal opportunities.

Today's parking order is not a result of scientific or rational considerations, but was fixed arbitrarily in the garage law of the Third Reich from 17 February 1939 (it became effective on 1 April 1939). It became practice in designation of areas and building development although building regulations do not adjust this. This data sheet shall help practitioners to substitute this practised usage with rules that are based on scientific principles. Therewith the regulations in the Austrian building regulations shall be introduced in terms of the goals of land-use planning.

2. Goals

The goal of this data sheet is to reach an equal treatment of the different modes

- Pedestrian
- Cyclists
- Public transport (PT)
- Individual motor car traffic

according to the higher-ranking goals of the Austrian traffic planning and land-use planning politics.

This can be reached through

- creation of a balance between modes
- creation of practical relevant, scientifically established recommendations for political implementation
- consideration of real mechanisms of action by the participation in transport

in urban and municipal areas.

3. Initial position

So far, increasing traffic and structural problems mostly are met with measures in flowing traffic, above all through the building of additional infrastructure. Especially with the introduction of compiuter modelling in transport planning and technology a focus on flowing traffic and its efficient handling can be noticed. With high expenditure symptoms are treated. This is coherent because problems in flowing traffic are noticeable by laypersons too. This approach results in urban sprawl, in the disintegration of urban structures, in the weakness and destruction of the commercial economy and the retail trade, in the outflow of the purchasing power of the historical city district, in traffic problems, in deficits in public transport and the manifold effects in the social, environmental and sanitary structure.

Already in the 1970s, recommendations of the [Organisation for Economic Co-operation] OECD ("Transport in Cities") refer to the essential control variable of the city traffic and the urban development, the "garage and parking management".

A problem of the conventional community and transport policies was that traffic was not organised as a complete system. Parts, coherent in reality, were optimised independently of each other, without considering the effects to the complete system. While parking space was optimised individually, public transport always had to be developed for the complete system. With this different approach no optimum for the complete system can be found.

4. General

The development of high and easily reached speeds through technical traffic systems is a very new technical innovation for which the human has no accordingly evolutionary equipment. The possibilities created through these new technical traffic systems were a fascinating perspective for the 19th and 20th centuries, particularly for individual user. The outcome of this system effect was not realised respectively ignored. Not until the second half of the 20th century the main causes of this behaviour could be explored and scientifically justified. It was demonstrated that to a less extent the flowing traffic but more the parking cars respectively the organisation of the parking space was the reason for scores of problems of the instability between the carriers. In this leaflet these scientific outcomes shall be combined for practical use.

Another source of this leaflet is the transport political requirement to integrate the public transport system according to its function on a competitive basis ("Priority for public transport"). The dominance of public transport disappeared with increasing motorisation, without concrete comprehensible essential causes.

To reach equal opportunity of the choice of means of transportation, structures have to be created that can truly offer equal opportunities. Primarily, equal opportunity acts in accordance with the indicators journey time (speed), comfort or rather the body energy input. Additionally there are other factors of comfort, accoutrement, cleanliness etc., which are however secondary factors. Taking into account the primary factor energy input there are quantitative minimum conditions, which have to be accomplished to leave open alternatives for the users. As the body energy input of the pedestrian per time unit amount – depending on speed – double to sextuple of the body energy input of driving a car, on the one hand the access to the public transport stop and on the other hand to the parking lot creates a physically measurable and comparable, which is objective factor for the choice of the means of transportation with a major impact on travel time.

The main starting point for the improvement of the legal realisation is the building order. Most Austrian building orders give a relatively wide scope concerning the spatial assignment of parking spaces to the structures for professional planners.

5. Comparability of public transport and car traffic

The parking space correlates according to the system effect with public transport stops. This is in no way expressed in the building rules. The organization of space cannot happen isolated from the higher-ranking objective.

Basic principle.

To assure equal opportunities, at least the footpath to the parked car has to be as long as the footpath to the stop of the public transport at several sources and destinations.

At given equidistance some footpaths become longer, like at the preparation of stops, some become shorter. The short footpaths to the garage will be compensated through a possible higher impact in the environment of the garage.

Therefore this equidistance to the parking space and to the stop is essential in land-use planning and in the organisation of space if one will grant a fair choice for the humans between these carriers – also to achieve the goals of land-use planning and to avoid the negative headed above.

6. Effects of the equidistance

6.1 For the citizens:
- Safe environment, as 70% of the street spaces become car-free and the living, working, leisure and contact spaces of the humans can be arranged qualitatively in a high-class way.
- Good environmental quality, as a big part of the living spaces will be relieved of car emissions and noise.
- High social integration capability of the community and the generations among each other as the street space is available again as communication space.

- Variety of numerous (but small) local shopping facilities.
- Comeback of workplaces and therewith employment opportunities in the living area.

6.2 **For the economy:**
- Fortification of the local economy revenues through the creation of local economic cycles with employment effects.
- Positive effects of employment, little use of resources, little effort for the motorised mobility.
- Enlargement and opening of new business fields.

6.3 **For the municipal administration:**
- Release of high maintenance and operating costs for road networks.
- Medium-term reduction of costs of the remaining infrastructure, as the settlements become more compact.
- Release of social charges, as social networks become sustainable and therewith informal social contributions can develop.
- Relief of communities through reduced payment of contribution for public transport, as it will be financially efficient with the higher number of passengers.
- Advantage for public transport: It needs fewer subsidies. Admittedly, there are more customers who are more critical, which is good for the system.

7. Financing topics in connection with parking

7.1 Basics
The current financing system in the parking area is counterproductive for a sustainable development of the structures. At present a countervailing charge is levied of those who not provide their parking spaces on their own plot or in their own house. Thereby, traffic, environmental and economic problems are produced, which burden the community.

Therefore the countervailing charge has to be disposed for parking spaces, which were not built. Instead there shall be a traffic exciter charge for those who park at home.

7.2 Introduction of a traffic exciter charge, dependent on the location of the parking space
Presently the principles of the market economy in the traffic system are not effective. So the users of the system receive false signals, which induce them to a behaviour that is against the objectives of land-use planning and transport policy, nor will it help them to adapt the structures in the complete system. Persons who behave falsely, are recompensed, those who behave right, are punished through the countervailing charge.

Persons, who park according to the structure, shall accomplish fewer charges than those, who create higher costs, thus park at home. The level of the charge acts in accordance with the particular local circumstances, with the current or planned supply of public transport and with the consequential costs, which emerge at present or in future in the system. For the calculation of the level of the charge the following function can be consulted:

Figure 10.1: Example of a cost function depending on the distance apartment/garage; shop/garage, workplace/garage etc. This function depends on the acceptance function according to the theoretical background, proved empirically in several cities: If the way to the stop is longer, the chance to use this footpath is lesser if you have alternatives. Analogously, the level of charge decreases with the length of the footpath to the individual vehicle (higher access resistance devotes less charges).

The charges shall be introduced according to the acceptance of the access widths – either floating or staged. The cost function has to follow the acceptance function. The cost function applies for the transition period until a balance condition is reached.

Example: A person, who parks at least as far away as the stop of the public transport, could pay as parking charge for the annual ticket of the public transport plus the operating costs of the garage. The person also gets an annual ticket for the public transport. A person, who parks at home, pays the costs, which he causes to the municipality or community. The surplus acts according to the privileges, which result of these real human behaviours. This price can amount the multiple of the costs of an annual ticket – depending on the distance to the stop of the public transport (Figure 10.1). This traffic exciter charge disappears, when the structure is stabilised again and all parking spaces are on the position, where they have to be in terms of a fair equal opportunity between the traffic modes.

The described differentiation is particularly necessary for the transition period until there is a balanced situation, then there are new finance structures possible and meaningful.

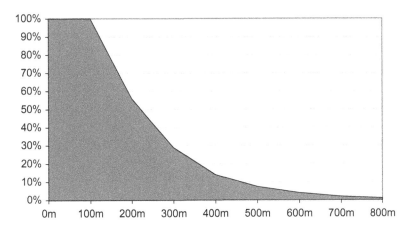

Figure 10.1 Cost function for the level of charge for near site parking.

7.3 Creation of transition periods

As the system is organised and operated constructionally, financially and organisationally contrary to the necessity of an effective transport infrastructure, a transition period has to be arranged, which has to be as short as possible. For the transition period today's distortion of competition, which exists between the peripheral shopping centres and the central shops in the city centre, has to be abolished. This can be reached when one has to levy a general parking charge (developing charge) at the shopping centres according to the number of parking spaces and the opening times, which accord to the short parking charges of the shops in inner-city areas. This charge needs to be adopted completely for the revitalisation and compensation adverse the inner-city shops.

8. Form of organisation

Generally it is to annotate that motor vehicle traffic is a technical system. But technical systems need an accordant organisation. Private operators (e.g.: in public transport, garage operators, control employments) of these organisations are common and possible.

Parking planning is not an individual affair. Parking planning and organisation are an exercise for the community and public. The provision and planning of all parking spaces has to happen through an organisation, which makes sure that the parking spaces are planned, built and managed like it is common with public transport stops. This has to be agreed with the objectives of the systems and the real behaviour of the humans. In this market economic system an equidistance of the parking spaces and public transport stops has to be demonstrated, in order to accord with the basic principles of the market. If one of these factors – proper building structure, proper financing structure and proper organisation structure – is going to be ignored, there is neither a solution for a sustainable city structure nor for the traffic problems.

At the moment, an individual organisation of the parking space is possible on own property. But the effect of the parking space in the system is equivalent to public transport in speed and in spatial impacts. Concerning the demand of the surface, the car exceeds the public transport multiple. Due to the same spatial impacts, the parking space cannot be organised individually, if the system should be stabilised. So the leaflet advises the establishment of an organisation, which is responsible for the allocation, organisation and financing of parking. It is also responsible for the financing of public transport and the rest of the carriers that means creation of a conformable traffic management at community level. Therewith the parking spaces will be organised on the right place to assure the financing according to balanced common finances of mobility. When someone needs a parking space or buys a car, they contact this organisation and get possible parking spaces and the appropriate price profile.

Important for the appreciation is that it is a necessary organisation of a technical system and not a bullying against the humans. Through this form of organisation, the human gets back the freedom of choice, which is taken today through the individualisation of parking.

9. Effects on the communities

With the introduction of an area-wide parking management the essential gap in the urban administration of resources is closed. Therewith, the community gets financial security for planning, also for public transport. The more, the community gets the responsibility for the essential control variable of the development of space and settlement and therewith makes an essential step to revitalise the city centres and cities.

In industry and trade equal opportunities will be established, as the high potential of shopping centres with nearly unrestricted supply of parking spaces also will be applied to market economic valuation. Well-organised communities with efficient short-distance traffic, short ways and favourable parking garages are attractive for industry and trade and achieve a liveable environment for humans, where a fair opportunity of choice is possible between carriers.

Prof. Hermann Knoflacher
Chairman of the Working Group "Verkehrsträger"
("Transport System Users")

Chapter 11

The changeable
shape of the city

Fabio Casiroli

CONTENTS

INTRODUCTION

"Whatever their main activity, cities still depend on being able to shift people and goods around with ease", as the great geographer Pierre George used to say.

It is no surprise, then, that urban studies can use some apparently unusual keys to deciphering cities, as in the case of the research presented in the "Mobility" section of the 10th Venice Biennale of Architecture. This research takes 12 megacities and analyses their territories using a standard area of 80 km by 80 km: Barcelona, Berlin, Bogotá, Istanbul, Johannesburg, London, Los Angeles, Mexico City, Milan, New York, São Paulo and Tokyo.

The main objective is to find a new way of understanding the urban organism, showing how its edges vary depending on the mode of transport and hour of the day, for each of a number of different ways in which the city can be "used". These have nothing to do, then, with the shapes we are used to seeing on city maps; those are defined by administrative boundaries, which may be a necessary simplification to ensure the urban machinery runs smoothly, but are inadequate for expressing the complexities or levels of use by residents and other city users.

The results obtained, the essence of which we set out here, are reassuring in the sense that it appears a method has indeed been developed that enables us not only to analyse this phenomenon, but also to measure the consequences of urban improvement measures that are indispensable for a better quality of the life in our biggest cities, these great conglomerations of people, functions and opportunities.

How will access to a given place change if a new metro line is built? How efficient will access to the road network be for a new multipurpose development? And how will particular contemplated measures fit into the existing city: how much better or worse will they make it? This method provides answers for these and other similar questions, provided we remember that this is just a first step, and its figures are bound to be approximate.

The changeable shapes of a city, then, are generated from the "focal points" of its occupation pattern and are transformed in accordance with the mode of transport used, as well as the time of day and day of the week that the city's facilities are occupied and used. This means that in order to read the shape of the functional city we need to use the fourth dimension: "time". It seems to us well worth looking into how an element as intangible by nature as time can become a real measurement tool for the city.

Our analytic approach uses a standard "mean time taken to get to 'use' the city": 45 minutes per journey each way, a value that has shown up in many studies and experiments, and one that allows each inhabitant to spend only a small part of their life on getting around. For looking at a truly swollen megacity, as we increasingly often are nowadays (in many countries the number of "extreme commuters" – those who spend over four hours a day in total getting to and from work – is increasing dramatically), we decided it would be useful to add a second isochronic band (45 to 90 minutes' journey time), for purely analytic purposes.

One instance will illustrate the whole arrangement: within the Milan megacity, the population of the potential catchment area from which the city centre can be reached within 45 minutes during the morning rush hour using public transport is twice as big as that reachable by private car within the same time (4.1 million against 2.2 million). Keep in mind that these are theoretical values, which do not correspond to actual effective demand and are quite different from the real capacity of the transport systems currently available: indeed, actual commuter journeys at these times are almost equally divided between private car and public transport. This gives, it seems, a sufficiently clear indicator to guide the allocation of resources that are available and can be activated (Figure 11.1)

When money is invested in public transport networks, it produces results in short order: the annual total of Tokyo's rail passengers has reached the incredible figure of 16 billion; 150 million passengers a year pass through the main metro station, Ikebukuro. Berlin's rail and tram systems attract 80% of public travel, 930 million journeys a year, and the rate of car ownership is modest: 322 vehicles per thousand inhabitants, compared with 758 in the Milan metropolitan area. Even in Bogotà, where getting around is not simple, the construction of the TransMilenio system has made it possible to carry 25% of passengers and to cut travel times by 32%, emissions by 40% and road traffic accidents along its paths by 90%.

The need to invest resources in improving public transport systems stands out even more starkly when we analyse the accessibility isochrones for stadiums, those pagan temples for sporting events. In the case of Milan on a Saturday evening, the population living within 45 minutes of San

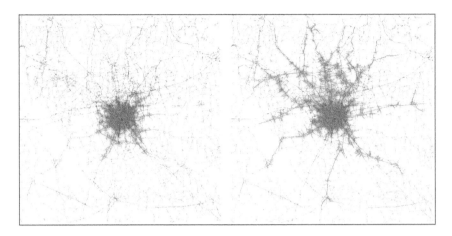

Figure 11.1 Milan: the morning rush hour in the working city, reachable in 45 minutes by private car (*left*), and by public transport (*right*).

Siro by car is nearly 11 times greater than the population that could get there within 45 minutes by public transport (4.3 million, against 400,000) (Figure 11.2)

In purely theoretical terms the shape of the city could be taken apart and put together again in an infinite number of configurations, though it could not, at its maximum spread, go beyond its outer boundary encompassing all users. Among these "possible cities" as delineated with use of the private car or public transport, limitations on space and visitor stamina have made it necessary to select some by deciding which functions

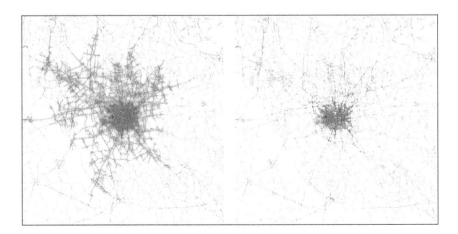

Figure 11.2 Milan: areas from which San Siro Stadium can be accessed in the evening within 45 minutes by car (*left*), and by public transport (*right*).

come first, and which times are of most importance, in terms of quantity and quality.

The "working city" comes first in this hypothetical ranking. It spreads out from the sites where the city's income-generating activities are concentrated and becomes manifest, with implacable constancy, in every weekday's morning timetable. Some of the megacities examined show, in the case of this particular time of day, situations of incredible efficiency and indeed manifest complexity: the public transport structures of Tokyo and London are remarkable; so are those of New York, Berlin, Milan and Barcelona; those of São Paulo, Bogotá and Mexico City are limited to the main corridors; while those of Los Angeles, Istanbul and Johannesburg are strikingly underdeveloped (Figure 11.3).

The second functional city we considered worth looking at was the "leisure city", the one whose centre of gravity and main attractors are recreational activities of various kinds (theatres, cinemas, public gatherings). This starts up and becomes visible, as a rule, in the first evening rush hour. Here we see the real strength of those megacities, which have well-established, integrated transport networks that are fully operational even outside working hours: London, Tokyo and Berlin. On the other hand, there are cities that have not yet devoted enough resources to these modes of transport, chief among them being Los Angeles. In that American megacity, the population living less than 45 minutes by car from Rodeo Drive is 27 times bigger than the population that could get there in 45 minutes by public transport (2.7 million, against 100,000): a clear case of social exclusion (Figure 11.4).

The third functional city is the "city of sport", which draws into its stadiums huge crowds of fans in the evening or on Sunday afternoons. Because these facilities are often sited near the edges of cities, we see an accentuation of the patterns already revealed by the "leisure city". In addition to London, Tokyo and Berlin, which we have already mentioned, Barcelona and Bogotá also show a well-spread shape for this particular function, with

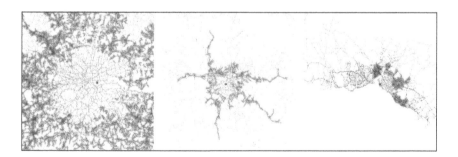

Figure 11.3 The "working city" using public transport. *From left to right*: London, São Paulo and Istanbul.

Figure 11.4 Los Angeles: the "leisure city" generated by the car (*left*), and by public transport (*right*).

large populations being well served (2.2 million out of 4.8 million in the former; 4.9 million out of 7.7 million in the latter) (Figure 11.5).

Fourth and last in this section comes the "cultural city", whose focal points are the great universities and that can draw in large crowds of students on weekdays, mainly in the morning rush hour. Columbia University in New York, the London School of Economics, Tokyo's Kishawa campus and the New Polytechnic of Milan can all be accessed from a very large territory within the standard rush hour times by public transport, in every case better than that offered by the private car (Figure 11.6).

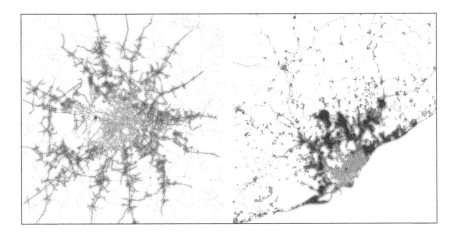

Figure 11.5 Accessibility by public transport (in the evening) of Berlin's Olympiastadion (*left*), and Barcelona's Nou Camp (*right*).

Figure 11.6 New York and Tokyo: accessibility of Columbia University and the Kishawa campus by public transport.

It seems clear from our account above that particular attention has been devoted to these public transport systems. The explanation is not to be sought in abstract ideological prejudice against the car, but in simple observation of the events that characterize city life. The private car, though an important instrument of individual freedom and a triumph that bears witness to a mature economy, has now shown itself to be the cause and effect of a devastating process of urban dispersion, as well as saturating a large portion of public space – which is a limited resource, very hard to regenerate, and that should be used as effectively as possible and enhanced by really innovative transport systems. The industry must raise its game significantly, and as soon as possible; perhaps by offering public transport arrangements that combine high speeds on long journeys with the ability to serve all the different parts of town with the same effectiveness. All cities, but especially megacities, must be regarded as complex living entities that do not just need embellishment with memorable architectural gestures, but also a thorough study of their circulatory systems and the ecology of their characteristic functions. Lastly, there has to be a major joint effort by academic experts and managers to plan harmonious development that takes proper account of internal relationships and relationships with the territory, and that offers proper urban living conditions to the enormous and ever denser population masses.

Research activity has not been limited to analysis, but, so far as it has proved possible to get information from the megacities investigated, has made efforts to assess prospects for the future. The readiest instance on which to try this scenario-painting is, for obvious reasons, Milan; but the exercise has also been carried out for Los Angeles, Bogotà and São Paulo.

In the case of Milan, where there is currently a lively discussion about the application of a London-style congestion charge, the results that emerged tend to strengthen the basic assumptions. The sequence of simulated schemes shows the potential of the new metro lines (some of which are already provided for in current planning instruments), and of the regional railway services.

The introduction of Metro Line 5 would make it possible to widen the potential catchment area of the "city of sport" from a population of 400,000 to one of 950,000; to offer a frequent, properly timetabled, regional railway service (SFR) even outside commuting times would increase that population to 3.7 million.

Similar results were obtained for the other megacities investigated (Figure 11.7).

We can derive certain conclusions from this research. It is possible to make cities more accessible at various times of day, provided we have policies for coherent measures that are properly based, above all else, on the construction or enhancement of the "network" public transport systems.

In circumstances where urban space is limited and expensive, great metropolitan areas (and also perhaps medium-sized cities) will no longer be able to allow free and unlimited road use, which produce gridlock, a steady worsening in the quality of life and a reduction in opportunities for socializing. It will not even be good enough to insist on the use of non-polluting private cars because, although they are an improvement for the environment, they change nothing in terms of taking up space.

Another definite aspect that must be kept firmly in mind is good town planning practice: intelligent siting of the primary urban functions, near to nodes that are better served by transport systems, will make cities more accessible to all, and therefore fairer.

The measures that do good are known; the tools for evaluating them properly exist; the resources can be found through a controlled involvement of private funding; and the general awareness that leads to claims for a right to mobility and health is growing everywhere. What is still missing in very many of the world's countries is a consolidated framework of rules for urban development; but what is even more vital to look for and bring

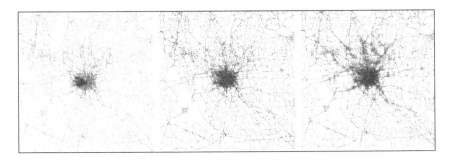

Figure 11.7 Accessibility of San Siro Stadium by public transport today (*left*), with the new metro Line 5 (*centre*), and with an improved railway system (*right*).

out is the courage, on the part of those who are responsible at various levels for governing our cities, to propose incisive measures rather than hide in facile consensus.

To make known our findings in the public forum is the duty of each and every one of us, both as experts and as citizens.

BIBLIOGRAPHY

Fabio Casiroli and Vincenzo Donato, *"Europe of Cities, Europe of Regions"*, 2006, CLUP (Milan Polytechnic Press).

Fernand Braudel, *"La Méditerranéè et le Monde méditerranéen à l'époque de Philippe II"*, 1949, Librairie Armand Colis.

Italo Calvino, *"Invisible Cities"*, 1993, Oscar Mondadori.

Lucio Stellario d'Angiolini, *"For a New Town Planning Praxis,1960–1990"*, 2005, CLUP (Milan Polytechnic Press).

Manuel Castells, *"City of Networks"*, 2004, Marsilio.

Mike Davies, *"Planet of Slums"*, 2006, Verso.

Pierre George, *"Précis de geographie urbaine"*, 1974, Presses Universitaires de France.

Saskia Sassen, *"The Global City: New York, London, Tokyo"*, 1991, Princeton University Press.

Chapter 12

Moving transport
Injecting transportation planning in Nairobi's metropolitan land-use agenda

Elliott Sclar and Julie Touber

CONTENTS

INTRODUCTION

Institutions matter

It is an obvious truism that good urban transportation systems are necessary for the solution of a wide range of urban problems. This is especially the case for cities in the developing world. The key to the creation of good systems lies in the process of decision making about the placement and timing of transport infrastructure investments and the quality of transport services. Transport systems, by expanding the options for residential, commercial and industrial location, hold the promise of permitting these cities to provide the means of absorbing the rapidly expanding populations that now inhabit these places. Cities in the developing world must quickly address the challenges of rapid population growth deriving from both nat-ural rates of local population increase and huge in-migrations from the domestic countryside as well as other countries. In this chapter, we seek to explore the potential for creating the necessary conditions for the type of decision-making process that can lead to the successful deployment and operation of good urban transport in cities in developing countries.

The stakes could not be higher. Besides improving local economic devel-opment options, local transport modal choices can contribute significantly to the improvement of the local and global physical environment. Transport systems that encourage pedestrian access, non-motorized transport and public transport foster cleaner air locally and mitigate global warming.

Good systems of urban transport enhance the ability of all urban ser-vices, but especially those related to public safety (police and fire protec-tion) and health (access to medical care facilities), to function optimally.

In all of these regards, the development of good urban transport should be carried forward with a special emphasis placed upon meeting the access and safety needs of the poor. Systems that improve access and safety for the urban poor invariably improve these for everyone else. The converse is rarely true.

The main challenge to success in the design and implementation of such improvements is not the comprehension of the technical requirements of transport network design or of the population's mobility needs. Rather, the critical factor is the ability of the social and political institutions in which the planning and implementation processes are embedded to man-age the enterprise. Thus transport research directed toward understanding and improving policy and planning implementation in general, and in the developing world in particular, has to be focused on critically analyzing the intersection between the institutional context of transport decision-making and the effectiveness of the resultant policy and planning agenda. When framed this way, institutional context is both a "given" in terms of constraints and a "variable" whose evolution must be considered as part of the transport planning process. A practical initial assumption for our work is a belief that while the quality of decision-making constrains what is possible at any moment in time, the engagement of decision makers with the substance of the problem, in this case land use and transport, creates real possibilities for improving the effectiveness of decision-making institu-tions. An important initial assumption from the standpoint of research is the belief that the institutions critical to decision making are susceptible to understanding and hence malleable to change even as the planning and implementation process itself proceeds.

In terms of research, two facts must be borne in mind. While the focus of such work is, of necessity, on the specifics of place and time, it is impor-tant to bear in mind that what is local is increasingly global. At the same time, understanding the local in depth is absolutely critical. In discussing the situation in a place like Nairobi, it is important to appreciate the ways in which its history and contemporary political and economic institutional arrangements define the specific context at any moment in time. In other words, history matters and path analysis is critical. Although the broad sweep of history is important to the shape of contemporary situations, in terms of the specifics of transport infrastructure in developing nations, the recent past looms exceptionally large, and is something akin to a dual-edged sword. On the one hand, this period has been marked by the emergence of new technologies with great potential for creating more opportunities to solve long-standing problems. On the other hand, these same technologies have often been used to constrain the options available to stakeholders in these places.

The transport decision-making institutions in nations such as Kenya reflect all of this. The heritage of Kenya's British colonial past is obvious in the present operation of its land markets and the formal structure of its political decision-making mechanisms. The history of colonization, because of the violence and the very particular racial forms that it took in Kenya, has shaped both the institutions of political decision-making and the organization of Kenyan society as a mixture of the modern and the traditional. This in turn has enormous implications for substantive forms of control over land, even when the formal aspects seem to clearly point in a different direction. Matters of land ownership and control have enormous and obvious implications for the process of designing and implementing land-use and transport policies for an emerging metropolis such as Nairobi.

Options in Kenya today are also a reflection of the fast-moving changes that characterized the post-colonial context. This context was initially distinguished by the paternalistic and cynical Cold War era of competitive African interventions by the Soviet State and the Western democracies, most notably the United States in the immediate post-colonial era from about 1960 to 1990. After the collapse of communism, there was a sudden policy shift away from active state intervention to policies of laissez-faire economic development among Western powers active in African development. This shift was most pronounced in the lending policies and structural-adjustment policies emanating from the international financial institutions. These changing policy decisions were absolutely critical for countries, such as Kenya, that are dependent on development assistance for infrastructure creation.

The new laissez-faire era, termed the Washington Consensus (Stiglitz 2002), was based on theories of economic development that significantly downplayed the importance of governance and hence the viability and effectiveness of social and political institutions. Such theories implicitly assumed that social and political institutions adjust to markets, and not vice versa. The theories under which international development policy proceeded assumed that there exists everywhere and always some rough approximation to the idealization of the standard-issue, competitive-market model that is the stock-in-trade of neoclassical economics. The goal of the Washington Consensus development policy was to set free this suppressed market so that it might serve as the central driver of economic development. Consensus advocates believed that this driver was so powerful that, once freed up, it would be capable of flattening the constraints that recalcitrant local social and political institutions might place in its path. But, as subsequent experience with this 1990s policy experiment quickly demonstrated, institutions are more powerful than markets. The markets that did emerge were a mere shadow of the idealized ones of market theory. Instead they resulted in bargaining arrangements shaped by extant social institutions and political powers, rather than the reverse. Therefore the key to influencing the effectiveness of transport and land-use decision making and its impact on social equity lies in understanding the connections between social and political institutions and the emergence of workable governance. Workable governance includes both market and non-market approaches to solving problems of economic efficiency and social equity. Williamson argues that, in the "economics of governance" (Williamson 2005), the key to economic development rests with the ability of the state to provide well-ordered and workable arrangements, especially for property structure. We place our concerns within this same framework. The implications for land-use patterns, and therefore infrastructure development, are thus linked to the state's capacity to organize the interaction between private and public actors.

It is not that the laissez-faire approaches to development ignored or overlooked the critical role of institutions, as much as they relegated the subject matter to either an impenetrable "black box" or assumed it not to be within the purview of their work. As Abramowitz has observed, among economists, institutions are "a measure of our ignorance" (quoted in Nelson 2006). And yet as Nelson and other thoughtful observers have indicated, institutions are the "basic rules of the game" (Nelson 2005). The challenge, then, for the development of an effective approach to development in general and to an applied field such as transport planning in particular, is to open the "black box" and identify and understand the set of institutions critical to success in creating a sustainable transport and land-use plan. In so doing, we can take the first step in the creation of the type of urban transport and land-use planning that will lead to sustainable urban development. We are seeking to uncover what we call the "conditions of positive policy change." It is important to understand that these are not static but highly dependent on time. The difference between success and failure may have less to do with the soundness of the analysis and much more to do with the timeliness of the accompanying action.

A necessary condition for positive policy change is the presence of a good networking process. A good networking process is one that involves governments and non-governmental organizations in dialogue and cooperation. To reiterate, good timing is a key element in this. A policy success story is usually the result of a strategy in which the series of moves or stages are prepared and anticipated, but it also relies on a conjuncture of events that can either help or abort this strategy. No matter how careful or well-executed the strategy, there are no guarantees. Success will always depend on single actors and single events that are beyond control. However, to the extent that the strategy is complimentary to and supportive of larger trends, it has a much greater chance of success. The current joint commitment to the Millennium Development Goals (MDGs), coupled with the media attention that the goals have raised, provides a good window of opportunity to promote positive policy change in African cities to meet the anticipated population demands of the coming decades.

DEFINING THE ECONOMIC ASSET OF A TRANSPORT SYSTEM

Urban public transport and economic development

A good public transportation system is more than the sum of its infrastructure and service providers. Good public transport systems share one important institutional characteristic: widespread acceptance by all the socially diverse populations that comprise contemporary cities. Acceptance requires that systems be safe, affordable, speedy, convenient and reliable. The success of global cities such as New York, London, Paris, Rome, Moscow and Tokyo rests in no small measure on the fact that they have public transport systems that meet these criteria. While none of the above cities gets a perfect score, it is still the case that all of them, along with many other similar systems in the cities of the Global North,[1] get sufficiently high marks that they contribute immeasurably to the well-being and prosperity of these cities.

The same generalization cannot be made for the cities of the Global South. Too many of the largest cities in middle- and low-income countries do not come close to meeting the criteria stated above. Instead, and especially in the poorest cities, public transport, to the extent that it exists at all, is largely an improvised system. Urban transport in these cities tends to be bifurcated and segregated by social class (Vasoncellos 2001). The most affluent residents are chauffeured or drive themselves in private vehicles on typically very poor road infrastructure in highly congested central business districts. The poor have little mobility. Walking is their characteristic mode of urban travel. They are often forced to share the same roadways where the more affluent ride. This lack of pedestrian accommodation contributes to both slower travel and higher rates of pedestrian fatalities. When the poor do use motorized transport, it is typically informal public transport in the

[1] "Global North" is a shorthand expression for the wealthier and more economically developed nations of the world that tend to be located in the northern hemisphere. "Global South" refers to the less developed and poor nations of the world that tend to be located in the southern hemisphere.

form of unsafe and often overcrowded minivans or aging taxis that charge prices that are comparatively high in relation to local incomes. Although travel times are somewhat better at the edges of the urban center, within it these conditions lead to chronic gridlock and travel speeds that are only slightly better than walking. The transport systems that serve the affluent are not public and the ones used by the poor are at best inadequate, improvised systems.

Although there is widespread agreement that cities in the Global South desperately need good urban public transport, such systems are not easily put in place. Ideas on what good systems might look like abound (ITDP 2006). There is no lack of transferable ideas to draw upon: for example, bus rapid transit, light rail, safe non-motorized transport, etc. The need for infrastructure and rolling stock financing is also not the principal bottleneck. On this measure there is a great deal of financing that could be accessed via international finance institutions, bilateral donors, local development banks, foreign direct investment and internal domestic resources. Rather, the principal obstacle is the belief that it is impossible to make anything of sufficient scale happen in these places. This pessimism can be traced to the, by now almost instinctual, belief that the formal and informal institutions of governance in developing countries are effectively dysfunctional. This in turn leads to a policy stalemate in which nothing can improve unless everything improves. It is this stalemate that must be addressed.

Because the problems associated with urbanization are now so pressing in these places—about one-third of the urban residents live in slums (UN Millennium Project 2005)—it is not good enough to argue that nothing can be done until effective institutions of governance somehow evolve. Nor is it good enough to just call for more aid on the grounds that the needs are dire. Instead it is important to engage these rapidly urbanizing places as they presently exist and begin to work simultaneously on problems such as creating effective urban transport and the process of institutional improvement. Progress on creating good urban transport will encourage improvement in governance, and improvements in governance will lead to further improvements in urban transport. Thus, this approach can be seen as a specific application of the more general approach to the challenge of economic development previously identified. It reframes the argument from one that we regard as a "chicken-or-egg debate" as to whether institutional change needs to come first (Easterly 2006) or large-scale investment needs to come first (Sachs 2005) into an argument that shows that these need not contradict one another.

Institutions and political economy as context

Central to the case presented here is analyzing and acting upon the crucial intersection between the formal and informal institutions that govern transport and land use in Nairobi and the actual workings of the system. This requires that we engage in institutional analysis. The method used here is economic analysis, which means that we evaluate institutions via their rationality and efficiency in serving desired social and private ends. This approach, labeled neo-institutional economics (North 1990) expands economic analysis beyond attempting to propose policies to facilitate market exchanges.

Rational institutional functioning then is akin to what Williamson calls the creation of "workable arrangements" or arrangements that connect social and private ends with alternative means (Williamson 2005). From an economic point of view the institutions that do this with the lowest possible transaction cost are the most desirable (Williamson 1975). Consistent with Williamson, Nelson (2006), North (1981) and others, we too hold that institutional evolution that rationally improves the connections between desired ends and available means is evolution that makes economic progress possible. It should of course be noted that we do not think that positive evolution is ever guaranteed. Institutions may move in an opposite direction. Indeed we will argue that the evolution of transport in many of the cities of the developing world exhibits precisely the perverse evolutionary style, moving from more efficient to less efficient forms.

From an evolutionary institutional point of view, it does not matter whether the institutional forms that emerge to solve the problem are market arrangements, public regulation or informal social arrangements. The key test for us is arrangements that effectively connect desired ends to means at the lowest possible transaction costs. Transaction costs include both the price and the non-market resources expended to achieve the desired end. Although in many instances, the most efficient, workable institutional arrangements can be created through market exchanges, these are not always the most effective structures of economic governance. When it comes to goods that have important social benefits as well as costs, as is the case for transport, other arrangements often work better. These include effective public regulation and public service provision, and, often importantly, in developing countries, informal structures organized around social roles and social status, in which exchanges are not done through market prices but rather through rules of social reciprocity. While we often term these latter solutions as "the informal economy," we would do better to call these institutions and transactions *the social economy* to distinguish them from the more familiar economy of market exchange (Lowenthal 1975). We seek to support a functional social economy to the extent that it supports broad-based social benefits and not narrow rent-seeking behavior.

The connections of these broader institutional concerns to urban public transport in developing countries are obvious and powerful. It is important to remember that public transportation and urban land use are two sides of the same coin. Hence questions about ownership and governance of urban land bear very directly on decisions about both the location and the effectiveness of any public transport system. Similarly given the informal nature of many existing public transport arrangements in cities in developing countries, it is critical that we analyze the institutions of this social economy in attempting to propose solutions that both improve governance and deliver important urban services. Hence we are seeking to broaden the capacity of the state to deliver public goods and services by more broadly conceptualizing the range of possible options for policy intervention. By widening the range of alternative interactions between private and public actors, we seek to direct the focus to the pragmatic creation of workable arrangements for governance of the complex transport land-use spatial arrangements required by these rapidly growing cities.

Transportation: a quasi-public good

Turning from a larger framing of the issue to the more specific question of transport, one of the major problems with market liberalization in the context of public services is that, while market liberalization is capable of allowing existing publicly run services to deteriorate, it is rarely able to create the market required to replace the public service with an equal, let alone privatized, alternative. This is especially true in the case of transport. As we shall see, in Nairobi, the preexisting public bus system was allowed to degenerate after market liberalization. In response, an informal transport sector emerged. It works well in some respects, but it is suboptimal in terms of the long-term urbanization needs of Nairobi and Kenya. However, because it does benefit a few key participants in that market, it gained an institutional life of its own. As a result, any move to improve the situation via renewed public sector activism cannot be accomplished by appealing to economic analysis. Success hinges on how well the political economics of policy change are fashioned and implicit recognition of the power of social institutions in this process.

Because of the unique features of urban transportation, this sector has been particularly negatively impacted by the efforts at privatization and deregulation that characterized laissez-faire development policy in recent decades. This is especially the case in cities in developing countries. As a result, despite that fact that populations in such cities are growing rapidly, there has been little improvement in these systems. The results are not only congestion and air pollution, but the concentration of large informal urban settlements close to urban centers to ensure that the poor can walk to employment in both the formal, but more often, the informal urban economy.

A particular problem faced by urban transport is a misunderstanding of its salient characteristics. From the point of view of the rider, urban transportation looks a great deal like an ordinary private good or service for sale. Individual riders can purchase transport trips in a manner that is identical to the manner in which they purchase cinema tickets or fruits and vegetables. Privatization and deregulation policy draws its analytic strength from this face of the service. If transport is nothing more than one more privately purchased good or service, the case for a passive role for govern-ment and a strong one for the unregulated market is strong. According to laissez-faire doctrine, the best that a government can do is simply get out of the way and permit the private market to supply the service in proportion to the intensity of the demand.

If the consumer side were all that is at stake in urban transport, this approach might make sense. But that is not the only side to the matter. Because of the power of urban transport to impact land-use patterns and environmental quality, the external impacts of the collective market-based decisions of travelers are far greater than the impacts on the individual traveler. Transportation powerfully organizes urban land use by the ways it permits and denies access to various groups in the urban population. We therefore define urban transport as a quasi-public good. Although it exhib-its the characteristics of a private good, its larger non-market or external impacts make it simultaneously a public good. Further complicating its sta-tus as a quasi-public good is the reality that it does not easily, if ever, cover its full costs via charges imposed on users. To the extent that we place more emphasis on the public characteristics of transport rather than the private, institutional analysis suggests that workable arrangements require that we look beyond market governance. We must approach the subject from the point of view of a public good with a minor role for market forces in deter-mining patterns of day-to-day use.

The ways in which urban transportation is subsidized are as important in determining its impact on urban life as the physical characteristics of a system. The need for perennial subsidy derives from the fact that trans-portation's fixed infrastructure costs for construction and maintenance of rights-of-way, i.e., roads and railbeds as well as the more variable costs of purchase and maintenance of rolling stock, can never be fully covered by charges to the base of users. Travelers simply stop traveling when the costs get too high, and then the ability of the area to capture the external benefits of good mobility are lost. Typically users are only charged a fare or toll that approximates the low marginal cost of their use plus a bit more toward the fixed costs of service operation. Thus, regardless of how the service is orga-nized, there is an abiding need for public subsidy and hence public policy and planning in determining how systems should operate.

UNDERSTANDING THE INTERACTION BETWEEN LAND USE AND TRANSPORT FROM A POLICY POINT OF VIEW

The two dynamics of metropolitan development

In the metropolitan development context, land use and transport represent two sides of a single coin. The combination of choices regarding housing is driven by this dynamic between land use and transport, and two types of dynamics have been identified in Nairobi: the satellite cities and the slums. This brings us to two kinds of congestion that define the emerging metropolis: a congestion of housing and a congestion of transport. In many ways they represent a trade-off with one another. Urbanizing Kenyans have to choose either to live close enough to walk to their places of employment, which means an unacceptable high density, or they have to move to peri-urban spaces, which means a difficult, expensive and often dangerous daily commute into the central business district.

These two types of congestion—traffic congestion and living congestion—reinforce the need for a comprehensive approach to the challenge of both types of congestion. In terms of scale, the city rapidly becomes too small a spatial unit. Planning at the level of the larger metropolitan region offers the only hope for resolving the stranglehold of the two types of congestion. In practical terms this means the creation of a planning process in which both the challenges of slum upgrading in the metropolitan center and physical planning for the surrounding satellite cities are part of the same process. For this to work, the communities of urban slum-dwellers, other Nairobi residents and residents of the surrounding satellite towns must be brought into a process of dialogue that recognizes the legitimacy of all of their various needs. In addition, all the agencies of local and national government as well as outside lenders and donors must be part of a legitimate planning process.

In concrete planning terms, we are seeking to create a balance that reflects the necessity of coherent infrastructure planning, a national concern, embedded within a participatory process that allows for local input within the metropolitan region. Clearly we are talking about issues of social institutions and political processes even as our targets are infrastructure and physical development plans. This in a nutshell is the challenge that Nairobi faces.

The green and brown divide: a heritage from the colonial history

Nairobi was established as the capital of Kenya by the British in 1907 and, from the very beginning, was developed with a clear spatial segregation between European settlements and African settlements (Mitullah 2003). That pattern has shaped the present social configuration of the city in terms of both income and living density. The western half of the city was reserved for Europeans, and Africans were relegated to the east. Kibera was an important and singular exception. Located on increasingly valuable land just 10 km from the city center, it is reported to be the largest urban slum in Africa, with a population estimated to be as high as one million people. Although Africans were not permitted to live in Nairobi during its colonial era, the British made an exception for Nubian tribesmen serving as soldiers for the colonial government. They were permitted to squat on that publicly owned land in the 1920s, as the British regarded them as politically more reliable than the Kikuyu, who would become the feared Mau Mau in the 1950s. After independence, because of the central location of this land, it quickly became a large slum settlement with an easy walking commute to most of the city's employment locations. The dominant ethnic group there now are Kikuyu. The arrangements of "landlords" and "tenants" there lead to very profitable and politically powerful arrangements that are not easily changed.

Land use was a central tool of social control for the colonial officials. At the moment of independence, influential Kenyan politicians and the rich took over this land-use control role (Olima 1997). The result of this history is that the notion of land ownership is closely linked to that of power and wealth and to social status within the society, and it became the nation's goal for its citizens.

There are two major factors that drive the physical development of the Nairobi metropolitan area, and both are inherited from the colonial era. First, the notion, as described earlier, that social identity is associated with land ownership, and thus the underdevelopment of a formal rental market. This has enormous implications for the ways in which the urban housing market emerged and it contributes to much of the tension surrounding land issues. Second, the relationship between the City of Nairobi and the satellite cities around Nairobi is based on the relationship originated during the colonial era, segregating the places where the Africans could live and where they could work.

In order to understand the very specific and profound interactions between transport and land use in Nairobi, it is important to remember that Nairobi was originally established as a city by the British in 1899 as a refueling point for the Kenya Uganda Railway (Mitullah 2003). The decision to move the headquarters of the Kenya Uganda Railway from Mombasa to Nairobi stimulated the growth of Nairobi, which quickly became a relatively large city by the third decade of the last century, with activities based around the railway industry (Mitullah 2003). Today, the railway company still owns 30 percent of the land in the city of Nairobi. This large property holding creates great opportunities for development, but it is important to understand that its roots are in the heritage of colonization.

As freight is more profitable than passenger rail, the recent privatization of the railroad company means that it will largely be used to transport freight from Uganda through Kenya to the port of Mombasa. Hence the newly privatized company will inevitably privilege freight over passengers, and the opportunity to use the rail infrastructure to guide the expansion of the metropolis via commuter rail will be lost. However, in this concession, the Rift Valleys Railways does not own all the land inherited from the Kenya Railways Corporation, as long as its use is not proved to be indispensable for exploitation of the rail (Concession Agreement 2006).

This leaves some flexibility to the government of Kenya to reassess the land and transport opportunity at the metropolitan level, opening up the possibility for some improvement in passenger rail transport, as the government can use this land control to create carrots and sticks to induce the private operators to provide for some form of passenger service.

Evaluating institutions: Nairobi transport

From the vantage point of the Global North, workable arrangements in the delivery of urban transport are typically viewed in terms of either publicly supplied or at least publicly contracted transport services. In cities in the developing world the provision of urban public transport is often a more complex amalgam. The situation in Nairobi is typical in this regard. Therefore in order to analyze the workable arrangements necessary to the supply of transport it is first necessary to disentangle the various elements of "physical and social technologies," the "institutional functioning," that constitute the effective transport system.

In Nairobi, at the most basic level, control of what Nelson would call the "physical technologies" is split between the public sector and the private sector. The public sector supplies the transport infrastructure through the Ministry of Roads and Public Works and the Nairobi City Council (Aligula et al. 2005). The private sector supplies the bulk of the transport services through fixed-route services provided by a private bus operator, City Hoppa, and the vast fleet of privately owned minibuses, known locally as *matatus*. There is also a publicly provided public transport service: the Kenya Bus Company. However, as a result of both a shortage of public subsidy and a turn away from the popularity of public service among policy theorists, the service level of the public service has fallen drastically. In the vacuum this disinvestment has created, the private operator, City Hoppa, and the *matatus* emerged to meet the demand.

One result of the stagnation of investment in public infrastructure is that the private sector has moved to fill the void in various places. The Nairobi Central Business Association has developed a number of pilot projects, such as the lighting system on the road between Kenyatta International Airport and the Nairobi Central Business (interview with Nabutola, September 2005). The cost is picked up by private revenues generated by advertisements on the light poles. Using this experience as an example, the business group argues that other similar private sector projects could emerge to address many infrastructure needs. While some elements of infrastructure supply that have commercial appeal could emerge, the larger problem of uniform improvement in transport infrastructure will invariably remain a public sector responsibility.

Although the challenge of adequate "private technology" is great, it pales in comparison to the need for what Nelson would call "social technologies" to make the organizational arrangements and interactions workable

(Nelson 2006). In Nairobi, several actors must be brought into the process of developing the organizational forms that will be needed if a viable and efficient system of urban public transport is to emerge. Because these actors are so numerous, making a comprehensive legislative framework operative is difficult. The actors include the Ministry of Transport, the Nairobi City Council, the Office of the President and the Kenya Revenue Authority within the Ministry of Finance (Aligula et al. 2005). Within these agencies there is a range of different individual actors that must also be brought into the process of planning and implementing any change in the transport system.

The strongest fact to emerge from an analysis of this institutional snapshot of the Nairobi urban transport context is that despite the interdependence of all these actors, there are no formal or, more importantly, ongoing institutional structures through which they can negotiate and cooperate. Moreover, because metropolitan Nairobi is spreading beyond the limits of the city into adjacent municipalities there needs to be some institutional context for more regional planning for transport services. Thus it becomes important that the Ministry of Local Governments and the Department of Physical Planning at the Ministry of Lands (Aligula et al. 2005), are also part of the transport planning and plan implementation process. There is presently a movement underway involving the Ministry of Lands and the Ministry of Local Governments to create a land-use planning process among the metropolitan municipalities. Clearly, if this is to succeed it will have to include the various actors involved in transport service provision and infrastructure development. Land-use planning cannot succeed without being closely tied to transport planning.

Looking at the window of opportunity

The formal institutional planning context within which all of the above take place is the "British"-inspired Metropolitan Plan. It is a reflection of British planning rules, as they developed during the colonial era. The 1973 plan is a complete master plan in the sense in which that term was used three to four decades ago. Despite its ideological heritage, and the events that have transpired in the intervening decades, both the Nairobi City Council and the Ministries of Lands and Transport, still refer to this plan, even if no part of this plan has been implemented as such. The 1973 Nairobi Metropolitan Growth Strategy technically expired in 2003. Institutionally this presents a good opportunity to rethink the metropolitan area, not only in physical terms but also in political and institutional terms.

In December 2005, the Department of Urban Planning at the Nairobi City Council and the Department of Planning at the Ministry of Lands began working on a concept paper for a new Nairobi Metropolitan Growth Strategy. The Permanent Secretaries of the Ministry of Lands and the Ministry of Local Government, including the Department of Urban

Development within the Ministry of Local Government, recently reviewed this concept note. This was the first step in officially launching the plan preparation that is intended to include a larger number of stakeholders.

The Nairobi Metropolitan Concept Note is important for three reasons. First, it sets the context for linking the infrastructure development, and so the economic development piece, to the institutional and political one. Second, metropolitan planning is an opportune tool to address social policy by thinking about redistribution in a very segregated context. Third, it provides a chance to reinvent a political process and institutions with a Kenyan identity.

CONCLUSION

Moving ahead: CSUD methodology

Based on the analysis of development and transport theories as well as the Nairobi contextual analysis presented in this paper, the Center for Sustainable Urban Development (CSUD) has developed a three phase project to reach the "conditions of positive policy change." This project is supported by the Volvo Research and Educational Foundations. To do this work, the CSUD has developed an integrated research and educational model based in part on the deployment of university resources and, importantly, upon maintaining its quasi-neutral status as a part of the academic world. This integrated model combines education, research and implementation with the goal of supporting local efforts at capacity-building, access to information and knowledge, and, in the specific case presented here, the definition of a uniquely African-defined urban policy. This model aims to contribute mainly to the training and accessibility of the new generation of planners and public health technicians in Africa. It also raises the question of redefining criteria for African cities and organizing a south–south axis based on similar experience.

Phase 1 of this project was completed in 2006. It was based on three components: an educational component, a research component and an implementation component. The implementation component of Phase 1 is described above. It relates to networking, or the identification of the local partners and "political champions" that will be able to carry out the project and thus the "conditions of positive policy change." The local partner here is the Department of Urban and Regional Planning at the University of Nairobi. This partnership was officially articulated in a memorandum of understanding signed in November 2005. The choice was directed by a desire to stay out of local politics and to assist in the training of the next generation of local planners. During the past two years, the CSUD has actively developed its broader local network, including central and local government actors, private sector actors, research centers/NGOs and media actors, as well as international institutions and funding agencies. In each category, the CSUD typically creates and articulates its local relationship through a letter of intent that formalizes both parties' understanding of the relationship. Strengthened by this network and its academic location, the CSUD has been able to organize a joint studio-workshop between Columbia University and the University of Nairobi—with a montage of interdisciplinary professors and students from urban planning, public health and international affairs. The added value of this initiative is hard to evaluate as it will hopefully last for an entire young generation as an exceptional experience. Most importantly, this initiative highlights a fundamental point: urban problems are now global, interconnected and multidisciplinary, and hence academic institutions must be global, interconnected and multidisciplinary in response if the challenge of the MDGs is to be met. Phase 1 research has focused principally on the development of white papers related to the connection between transport infrastructure and public health.

Phase 2 of this project is ongoing and is defined by two initiatives: a first one directly linked to implementation, the Ruiru Initiative, and a second one based on a research and education effort intended to impact the policy-making decision process. The Ruiru Initiative can be summarized as the construction of a physical development plan for a satellite city for Nairobi. The goal of highlighting a satellite city is to assist local efforts at creating a high-priority agenda for metropolitan planning. The effort in Ruiru has permitted us to assist in strengthening the links between the different administrative units at the national and local level in a regional planning effort. Transport becomes an obvious and powerful connection once the region becomes the focus of a planning effort. The Ruiru Initiative consists of five steps that overlap different phases of the Nairobi Project. Step 1 was setting up the network framework to move the initiative forward. We did this by building an official partnership with the Ruiru Municipal Council and the Ministry of Lands (local and central governments). This led to a consultation process with all the stakeholders at the metropolitan scale in order to identify the needs. Step 2 was the joint studios-workshops previously described. Step 3 involves transforming the student work into formats suitable for presentation to local and central governments, as well as residents and local landowners, who need to be a continual part of the participatory process. Step 4 follows directly. It is the processing of this work and analysis to the community. It is difficult to explain how valuable this has been as it involves an ongoing trade-off between the capacity of professional analysis and the priceless local knowledge, which is usually not recorded but that must be brought on bear on revisions to any planning effort. Step 5 will be the creation of a set of three plans, a strategic development plan, an implementation plan and a financing plan. These plans will be critically shaped by the participatory planning process that includes the different stakeholders and will highlight original financing package ideas.

The second initiative of Phase 2 is to support the development of local research capacity in transport and public health (air quality focus). These two research/educational initiatives are conducted on the same model: the CSUD has just established partnerships with two non-university research institutions, the Kenya Institute for Public Policy Research and Analysis (KIPPRA) and the Regional Center for Mapping of Resources for Development (RCMRD). Both institutions have established their relationship with the CSUD as a same level exchange of data access and savoir-faire, the use of software for modeling or monitoring for example. The goal here is to create a sophisticated transport model and an air quality monitoring system at the Nairobi metropolitan level. During the completion of this effort, training seminars will be organized in order to spread this savoir-faire so that it can be duplicated or even expanded at the East African level later on.

Phase 3 is implementation. Phases 1 and 2 emphasized the process of planning and set up the conditions for a policy change. We took great care carrying through Phases 1 and 2, especially the inclusion of many local partners in the research/educational/implementation combination so that the ownership of the project would be local and one that is accepted by all stakeholders. This is still a work in progress and will be evaluated at the completion of the project and even well after. But in order to conduct the type of evaluation that will be needed, it is necessary that we document the process at each step along the way. We believe that our work will contribute positively to the larger discussion of effective approaches to international development and planning. There are many successful small-scale projects in Africa and around the world, but the real issue of development is to be able to scale up these small successes to make a measurable impact at the level of the city and the region. Only the institutions of a country can make this a reality. The real challenge is thus to bring these institutions into the debate. Transport infrastructure presents so many opportunities at the urban scale that it provides a wonderful starting point.

ACKNOWLEDGMENTS

Celeste Alexander and Jennifer Schumacher-Kocik

REFERENCES

Aligula, E., Abiero-Gairy, Z., Mutua, J., Owegi, F., Osengo, C., Olela, R., 2005. *Urban Public Transport Patterns in Kenya: A Case Study of Nairobi City, Survey Report, Special Report No 7*. Kenya Institute for Public Policy Research and Analysis, Nairobi, Kenya.

Easterly, W.R., 2006. *The White Man's Burden: Why the West's Efforts to Aid the Rest Have Done so Much and so Little Good*. Penguin Press, New York.

Institute for Transportation and Development Policy, 2006. Sustainable Transport: A Sourcebook for Policy-Makers in Developing Cities. www.itdp.org.

Interview with Nabutola (by Julie Touber), September 2005. (Unpublished). Nairobi, Kenya.

Lowenthal, M., 1975. The Social Economy of the Urban Working Class. In: *The Social Economy of Cities*, eds. Gary Gappert and Harold M. Rose. Sage, Beverly Hills, CA. pp. 447–469.

Mitullah, W., 2003. Urban Slum Reports: The Case of Nairobi Kenya. In: *Understanding Slums Case Studies for the Global Report on Human Settlements*. UN Habitat, Nairobi, Kenya. pp. 195–228.

Nelson, R.R., 2005. *The Limits of Market Organization*. Russell Sage Foundation, New York.

Nelson, R.R., 2006. *What Makes an Economy Productive and Progressive? What Are the Needed Institutions?* LEM Working Paper Series. Laboratory of Economics and Management Sant'Anna School of Advanced Studies, Pisa, Italy.

North, D.C., 1990. *Institutions, Institutional Change, and Economic Performance*. Cambridge University Press, New York.

North, D.C., 1981. *Structure and Change in Economic History*. Norton, New York.

Olima, W.H.A., 1997. The Conflicts Shortcomings and Implications of the Urban Land Management System in Kenya, *Habitat International*, 3, 319–331.

Republic of Kenya, 2006. The Kenya Railways Corporation and the Government of Kenya and Rift Valley Railways (Kenya) Limited. Agreement Providing for the Concession of Kenya Railway Freight and Passenger Services. Republic of Kenya.

Sachs, J., 2005. *The End of Poverty: Economic Possibilities for Our Time*. Penguin Press, New York.

Stiglitz, J.E., 2002. *Globalization and Its Discontents*. W.W. Norton, New York.

UN Millennium Project, 2005. unmillenniumproject.org.

Vasoncellos, E., 2001. *Urban Transport, Environment and Equity: The Case for Developing Countries*. Earthscan, London.

Williamson, O.E., 1975. *Markets and Hierarchies: Analysis and Antitrust Implications*. Free Press, New York.

Williamson, O.E., 2005. The Economics of Governance, *The American Economic Review*, 2 (95), 1–18.

Chapter 13

Urban public transport and economic development

Harry T. Dimitriou

CONTENTS

INTRODUCTION

The excellent presentation to which this chapter is a response describes "the process through which transport policy considerations are being introduced into the local land-use agenda" of Nairobi in Kenya.

It identifies and articulates the necessary conditions considered by the authors for a "successful" transport and land-use policy to elevate social equity to a higher priority of economic development.

It further argues that metropolitan institutional mechanisms for land-use planning provide the most appropriate "policy space" in which to articulate a complimentary, sustainable, public transport policy.

The authors begin the chapter by arguing that it is a truism that "good" urban transportation systems are necessary for the "solution" of a wide range of urban problems. This claim has, however, two embedded premises that warrant further examination:

The first is that the concept of what is "good" is a constant and universally agreed.
The second is that the idea of arriving at "solutions" (rather than moving towards "better" situations) is indeed possible.

The fact of the matter is that the vision of sustainability has dramatically altered what the experts previously considered were "good" transport systems. No longer are concerns of transport operational efficiency "king" over all else.

If examined carefully, the concept of "solutions" is linked to preconceived "obstacles" to the achievement of specific goals/targets held by particular parties, stakeholders or vested interests as the main criteria for judging success and/or failure.

More importantly, not only do many of these criteria conflict, they are also ultimately linked to agendas that go well beyond the concerns of local development.

The question, then, that emerges is which agenda is to be given priority (and why)? And what to do first, invest in public transport in a manner that generates wealth and productivity or in a manner that enhances equity and reduces poverty? "Win-win" strategies for achieving both simultaneously are much more difficult to implement, especially where private sector investment is critical and where short-term monetary targets prevail.

CREATING THE NECESSARY CONDITIONS FOR EFFECTIVE DECISION-MAKING

There is no doubt that the development circumstances of places such as Nairobi make the stakes on almost all fronts very high indeed – both for the poor and the rich. This in turn leads to desperation among the underprivileged, opens up more opportunities for the exploitation of the poor and increases corruption practices in both the public and private sectors.

Under these circumstances, low-cost transportation investments that provide greater access to education, health and employment opportunities, and therefore offer potential "hope" to the underprivileged can only be welcome.

It is then very puzzling why, throughout the developing world, the international development banks have invested so little in the "lowerarchy" of transport infrastructure, given its potential to support non-motorised movement to key opportunity areas, so important for the urban poor. Could it possibly have something to do with the limited rates of return such projects offer these banks, suggesting that there is a rhetoric in their international reports not matched by the reality of their project portfolios?

The rhetoric of the international development banks, however, is only part of the overall malaise. For it is more than matched by the rhetoric of the politicians, who too often promise a motorised vision of the urban future, which is not only unaffordable in terms of the amount of infrastructure required but also unaffordable (and unobtainable) in terms of the scale of energy levels needed to fuel such a vision.

As if these handicaps were not enough for cities in the developing world, we have the "tyranny" of the experts – the city and transport planners

– obsessed with concepts of transport operational efficiency above all else and with concepts of "orderly" land-use patterns that are not only unachievable but undesirable. Such concepts have been reinforced by much education and training, the academic and professional literature they rely on and their professional visions of development and "order".

This state of affairs has much to do with the bankers, politicians and too many professionals ignoring the "wisdom of crowds". As one rickshaw driver once explained to me with great frustration in the 1990s, when I was conducting a travel survey in Kanpur, "If the transport planners really wanted to find out what the needs of the transport poor are in Indian cities, all they have to do is to ask the rickshaw drivers, after all, we are among the most underprivileged in Indian cities and know a great deal about the movement needs of others".

CONTEXT AND HISTORY MATTERS

The authors make mention of the importance of "the specifics of place and time" and ask "to bear in mind what is increasingly global" needs to be set against an in-depth understanding of what is important locally. This is most important.

They highlight the significance of understanding the impact of a place's history on contemporary political and economic institutional arrangements, and the "path dependencies" created for physical infrastructure provision and institutional governance.

In the presentation, one notes the highlighting of the collision of technology and infrastructure heritages that hark back to the British colonial past with contemporary infrastructure needs and the technological changes bringing to Nairobi an increasingly globalised future. This is a future that is dramatically affected by fast changing information and transport technologies that, we are told, have the power to prevent the city from being "by-passed" by the global economy.

The American business model exported

The laissez-faire model (termed "the Washington consensus" by Stiglitz) promoted by international agencies such as the World Trade Organization (WTO), the International Bank for Reconstruction and Development (IBRD) and the International Finance Corporation (IFC) – which I refer to here as the American business model (ABM) – seriously downplays the institutional governance dimension of development and the significance of local engines of change in favour of an "idealized global competitive market", more virtual than real, described by some as being more of an ideology.

In the last part of the twentieth century and the first part of the twenty-first, setting markets against institutions and denigrating ideas of governmental intervention (as a means of correcting faulty markets) in favour of the acclaimed merits of the free market, have fragmented communities, cities, city-regions and even nations. They have done this by, among other things, promoting infrastructure and services that make places, communities and economies increasingly compete rather than integrate, setting the "haves" against the "have-nots" and creating new "winners" and new "losers" in a more uncertain and unsustainable future.

URBAN PUBLIC TRANSPORT AND ECONOMIC DEVELOPMENT

Any attempt to confine urban public transport performance appraisal to fiscal rather than welfare concerns *as well* is, together with the privatisation of public transport services without subsidy support for the urban poor or broader development concerns, nothing short of an ideological move.

The penalties of failing to recognise this can be immense, especially in cities where the majority of inhabitants live below the poverty line or where a major section of the residents are poor. Providing profitable and efficient services along selected routes *only* to leave the "rest" to their own devices – or, worse still, to the illegal sector – is to fail in innovative, public-focused thinking that could see the introduction of more affordable public transport services in collaboration with the informal sector.

Looking to the cities of the "so called" developed world for standardised "solutions" for public transport provision for cities of the "developing world" – be they bus rapid transit (BRT), light rail, metro and/or non-motorised systems – has three shortcomings:

- It encourages "silo thinking" and sets advocates of one mode against another, rather than looking at the merits of each and building up an integrated multimodal system, where each mode is used to its optimum in the light of the energy, financing and capacity advantages it offers.
- It takes on board (albeit perhaps unintentionally) the transport path-dependency and vision of the overseas city, from which the systems model is mimicked, rather than closely examine local needs as a priority.
- It fails to take into account the management and institutional capacity prerequisites for such systems to succeed, which may or may not be present in the city of the developing world.

The need to fuse concepts of "formal order" and "informal order" at the appropriate time in order to optimise opportunities and benefits is very important.

In their efforts to become economically sustainable in an increasingly globalised and uncertain world, global corporations such as IBM have of late learned that potentially the most innovative thinking takes place within less structured and smaller units of the economy, and that, notwithstanding their size and informal nature, these units can prove critically important to the success of the larger highly structured units.

This mode of thinking, I would advocate, is most suited to land-use planning and urban management, multimodal public transport planning, and integrated land-use/transport planning in cities of both the developing world and the developed world, especially the former.

With this perspective in mind, the "tyranny" of the transport experts – imposing highly structured, motorised transport systems on very unstructured urban land-use patterns – could be seen to represent the imposition of a particular highly formal order onto an extensive informal one that, ultimately, it displaces or overrides.

What is ironic is that this newly introduced order rarely leads to the idealised overall structure the city and transport planners envisage in their master plans, but instead a new "emergent order" develops that represents almost an organic attempt by the city to adapt to the new reality.

The premise I'd like to pose here is that if we were better able to understand the ingredients and dynamics of this "emergent order", we would become better planners and be more proficient in providing enhanced and more appropriate housing, transportation and other city development services. Significantly, this way of thinking flies in the face of the idea that "nothing can improve unless *everything* improves".

UNDERSTANDING THE INTERACTION BETWEEN LAND USE AND TRANSPORT

The interaction of land use and transport has long been recognised to be a complex relationship. Despite the pretence of some acclaimed simulation efforts in modelling, and the rhetoric of the planners in their claims to be able to integrate land-use planning with transportation planning, the reality is that it remains a poorly understood relationship and, therefore, an inadequately managed or planned one in cities of both the developed and developing world – especially the latter.

If this relationship was considered complex and unmanageable before, then the rise in importance of the third dimension – the environment – can only make it "ultra complex". What the climate change and carbon emission concerns associated with cities and their transportation activity have done is to make how we tackle our urban movement problems locally highly significant globally, and, in so doing, they have placed a great deal more pressure on public transport systems and public transport providers to deliver more sustainable solutions.

The fact that the global externality costs of motorised urban transport have become irrefutable worldwide – as has the need for international, national and local agencies and governments to introduce intervention measures in an effort to bring trends within manageable levels – should make it easier at the local level for externality concerns to now be incorporated within a broader formula of deciding what is a successful public transport system.

However, the simultaneous greater reliance on private sector capital to build, operate and finance fast-rising urban transport infrastructure and service needs, especially major urban public transport projects, not only opens the door to increased privatisation and raised expectations by global investors for quick, high financial returns, but also reveals their unwillingness to take on long-term risks without government guarantees.

The call for more sustainable futures, cities, transport systems and lifestyles promises to put these visions on a collision course with that of the current ideology of globalisation – based as it is on competitiveness (rather than collaboration) and on ABM.

This ABM rationale prevails among the growing army of global infrastructure investors and public transport providers who are at present typically "regulated" by weak public sector institutional agencies too coy to challenge the exploitative PFI/PPP/BOT practices of such companies for fear that they (the global investors) may not invest in their projects but in a competitor's instead.

What these circumstances highlight is the need for a strong and competent public sector to enable a vibrant and competent private sector to prosper. (Something even *The Economist* advocates.) The absence of strong public sector institutions must be the Achilles heel of all such cities as Nairobi if they are looking to attract effective global investment. This concern is quite rightly one of the major focal points of the CSUD methodologies proposed by the authors.

Chapter 14

Urban mobility in China
Developments in the past 20 years

Haixiao Pan

CONTENTS

MOBILITY AND URBAN DEVELOPMENT

Urban areas play very important roles in economic and social development, and with more than 20 years of rapid economic growth, urban development has achieved remarkable success in China. In a large city the advantages of a better environment for investment and economic conditions, well-equipped city functions and relatively high input-output efficiency are especially notable. To create wealth in the cities, many activities have become increasingly frequent, whether they are based around work and business, leisure activities, or visiting friends and other social activities, and so on. Intracity and intercity linkage have become closer. The existence of the city itself is largely dependent on the function of facilitating communications. Such contact is bound to be accompanied by the movement of persons and goods. "Mobility has become the most fundamental value of a society, achieving social change, the prerequisite for development and progress,"[1] as the French scholar François Ascher said. Mobility reflects the ability to bring about movement despite the constraints of various obstacles or barriers, and is now seen as a "fundamental right " for humans; it is the prerequisite for people to access their other rights of working, living, education and health.

The vitality of a city is largely due to the diversification of activities to participate in and which contribute to the city. At a basic level of urban mobility, people will first think of traffic congestion, and longer time to travel, before they become involved in their various urban activities. This will inhibit people's desire to come into contact with other people, limiting the potential for urban development. Only when urban mobility reaches a certain level, will people be able to join more diversified activities, in order to create the unique style of a city[2] (Figure 14.1).

Following the principle of sustainable development, an urban transport system should provide the opportunity for more convenient and more efficient movement of people and goods.

The transport infrastructure provides the physical guarantee for people and goods to move, and it also has the function of a catalyst to induce new socioeconomic activities. Poor urban transport services will inhibit the movement of people, restricting urban development. For several decades after liberation, major cities in China depended heavily on travelling by foot and by bicycle, and basic services and facilities were located nearby: the "big yard" layout of the workplace and apartments helped people to work within walking distance. For many years in the planned economy, the low standard of living framed a compatible urban spatial structure with a

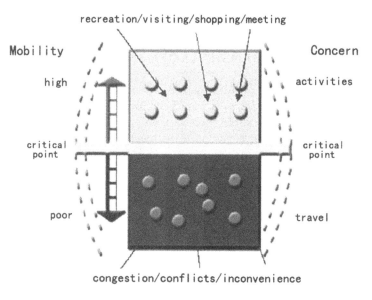

Figure 14.1 Critical point of mobility and activity.

slow transport mode. At that time, people belonged to various companies and institutes. The desire to move had been constrained to the lowest level for socioeconomic linkage. The major city itself was actually degraded as an isolated "industry village." Moreover, due to the state policy of encouraging heavy industry in remote rural-mountain areas for defense, there was a long delay in the construction of urban infrastructure. However, traffic conflicts were not prominent. For example, the public transport in the 1970s operated at an average speed as high as 30 km/hr.

After the reform and opening up of China, this balance was soon broken, and with long-term debts in transport infrastructure construction, the urban mobility level is very low. One of the most prominent problems is a serious shortage of supply of the cities' road systems. In Shanghai's central area, for example, the road area per capita was only 2.29 sq m.[3] Another feature is the huge volume of cyclists on the streets. In some megacities, during peak hours, the bicycle traffic volume was more than 10,000 on major roads. Poorly developed city road systems and a difficult and conflicting combination of motorized and non-motorized traffic competing for very limited road space resulted in very low level of mobility. In the early 1980s, it would take more than two hours from People's Square to Shanghai Hongqiao airport – which now less than one hour, even though there has been a tremendous increase in the number of motorized vehicles. In order to guarantee the people's basic needs of mobility under the constraints of very limited resources, "limited stop" express bus and "mother/child only" bus services were provided. So, for a long time people believed that the key path to improve urban mobility was to improve the urban road network and increase road capacity. The second change to the second important policy is how to promote bicycle travel being transferred to travel by public transport.

THE MAIN FEATURES OF THE DEVELOPMENT OF URBAN MOBILITY

Since the 1980s, Tianjin, Beijing, Shanghai, Guangzhou and other major cities in China have been carrying out a large-scale survey of urban transport and urban transport planning. Based on this work, along with the rapid expansion of urban areas, large-scale urban transport construction has been taking place. The major measure to improvements in urban mobility is the intensive construction of road systems, with the following aspects as important features.

1. High intensity of investment

   For more than 20 years, a high intensity transport investment has been in place in large cities in China. Beijing's investment in the transport infrastructure in the "Ninth Five-Year Plan" period is RMB 40 billion yuan, accounting for 4.3% of the gross national product (GDP). In the "Tenth Five-Year Plan" period this is expected to reach 83.8 billion yuan, 5.15% of GDP. In Shanghai, before the "Seventh Five-Year Plan", in the past 37 years the total investment in urban road transport was about 15 million yuan. Investment in 2003 amounted to 22.1 billion yuan. For Guangzhou, in 2002 and 2003, the investment in urban transport accounted for 3.5% and 3.2% of GDP respectively.

2. Remarkable increase in urban road space

By the end of 2002, there was 191,000 km of urban roads, or a total of 276,361.89 million sq m of road space for cities in the Chinese mainland. The per capita road space rose from 2.8 sq m in 1980 to 7.8 sq m.[4] In Beijing the road space was 21.45 million sq m in 1986, reaching 43 million sq m in 2003. In Shanghai, from 2002 to 2003, the road space increased by 10%.

FOCUSING ON RING ROAD AND EXPRESSWAY CONSTRUCTION

Due to the lack of high-grade roads and the lack of capacity to carry significant volumes of traffic, all cities focused on the construction of high-grade road systems. In Beijing, since the 1980s, the principle of extending the city on the east/west wings to decentralize the central city has been accepted and the ring road strategy was implemented, with five ring roads. The fifth ring road, at 98.8 km long, cost 13 billion yuan.

In Guangzhou, ground transport has been basically formed with three level ring road systems (inner ring road, city expressway, Huanan express trunk road, and the north second ring road). Based on the ring roads, there are 25 radial roads from the inside outward and 45 regional roads, constituting the main backbone of the road network system.

After eight years, Shanghai has finally built a "申"-shaped system of elevated motorway as well as three west–east, north–south city trunk roads. In 2003, there were 44 lanes connecting both sides of the Huangpu River. It took ten years for Shanghai to complete the outer ring road with a total length of 99 km, and the total investment in the road was 17.544 billion yuan. These high-grade roads have a very significant role in carrying the traffic in the city, accounting for 20% of the total length of roads in Shanghai center, but accommodating nearly 70% of the traffic.

Each traffic lane in Beijing's ring road will take as many as 1600–2100 passenger car units per hour (pcu/hr), almost its saturation capacity. However, the effects of improving the city's operating efficiency are not obvious. Currently about 40% of people in Beijing spend more than an hour travelling to work each day.

ENCOURAGING ATTENTION FOR PUBLIC TRANSPORT DEVELOPMENT

Encouraging public transport development is a basic urban transport policy. To break the prevalent "vicious cycle" in major cities – "more road, faster increase in the number of vehicles, more congestion" – in the first half of 2004, the Ministry of Construction promoted the "priority on the development of urban public transport". It clearly sets forth the principles of "giving priority to the development of urban public transport strategy". China will strive for a period of about five years to establish public transport as the dominant mode of transport in urban areas.

In China's large cities, we have gradually established conventional urban public transport networks. In 2003, all Chinese cities combined had 259,000 standard buses, a 9.3% increase over the previous year, 7.66 buses per 10,000 people. In the same year, Shanghai had as many as 18,625 buses and 952 operation lines.

Public transport is at a disadvantage compared with conventional cars in terms of comfort, speed and reliability. While we have tried to expand the urban public transport service, it is still far from meeting urban transport planning strategy goals, even lower than the modal share of the mid-1980s. In 1986, for all forms of mechanized modes, public transport accounted for 32% of traffic. It reduced to 26.5% in 2000.

Improving the operation speed of public transport is essential. In 1999, Kunming launched a bus-only route, giving an increase in bus speed from 10 km/hr to 15 km/hr. From 1999, after the bus priority strategy had been established in Kunming, the number of daily passengers increased from 400,000 to 1 million and the bus share increased from 8% to approximately 14%.

In 1997, Beijing opened on its first bus lane along Changan Avenue. Currently, the total length of bus lanes is now greater than 100 km. To further improve the urban public transportation operating environment, Curitiba has been taken as a model to follow for constructing a high-capacity bus rapid transit (BRT) system. On December 25, 2004, China's first BRT was been put into operation in Beijing.

RAPID DEVELOPMENT OF TAXIS

From 1993 to 2003, the number of taxis in China soared from 190,000 to 884,000 vehicles, with an average annual growth of 16.6%. In total, the taxi industry employs about 2 million people. Beijing, which had nearly 60,000 taxi vehicles in 2001, 13.6% of the ground public transport passenger total is borne by taxi. The annual passenger volume reached 630 million people by taxi in 1986, accounting for less than 1% of all travel, and rose to

8.8% in 2000, due to the convenient service and relatively low prices. The use of taxis became an important mode of transport in daily life, in particular for high-level business, leisure and recreational trips. Taxis can help the city government's revenue, meeting the demands of door-to-door travel, and they are a comfortable and convenient flexible, personalized transport mode. Providing a high-quality taxi service, combined with parking controls and license bidding, will inhibit the demand for family cars, which will also contribute to improvements in urban transportation efficiency and urban mobility.

CONSTRUCTION OF LARGE-SCALE METRO SYSTEM HAS BEEN WIDELY ACCEPTED

The scale of population and building in China's cities is very high. According to the general principle between urban density and urban transport mode choice, rail transit is the inevitable choice for transport strategy in these large cities. By the end of February 2004, a total length of 292 km rail transit had been completed and put into operation in Beijing, Shanghai, Guangzhou, Dalian and Tianjin cities. The central government has approved 11 urban rail transit projects for 320 km of new rail transit systems, with an investment of 102.5 billion. More than 10 cities have completed the urban rail transport network planning. In the next 10 years, Guangzhou rail transit network will extend to a total length of 537.5 km. All the revenue from land leasing in Guangzhou will be used for the construction of the metro system.

After 10 years of construction, Shanghai has completed and put into operation rail transit lines 1, 2 and 3 with a daily passenger volume of 1.3 million people, which accounted for about 10% of the total public transport passengers. By 2010, Shanghai intended to have 12 rail transit lines. And by 2020, the total length of rail transit will be 810 km, of which 480 km will be in the central city.

THE CONSTRUCTION OF A HIGH STANDARD, INTEGRATED TRANSPORT HUB

The transport hub is a distribution center for the flow of passengers and vehicle traffic, where various transport modes can be gathered in one place in order to provide a choice of modes, so that those traveling can fully use the advantages of each mode of transport.

A convenient and well-integrated transport hub which takes account of the needs of people will make travel between cities and within a city more comfortable, reliable and time-saving. Without an efficient transport hub, the efficiency of a city transport system will be decreased.

In the past, when planning urban development, much less consideration has been given to transport hubs, meaning that each mode of transport is not well integrated, and this also makes public transport less attractive. Recently, however, the construction of transport hubs has been paid special attention. Learning from the concept of a "seamless" transport system, Shanghai will have more than ten large-scale transport hubs (for example, Xujiahui, Wujiaochang, South Railway Station, etc.) in order to help realize the so-called "zero distance transfer".

In addition, because of the high volume of people, activities can be grouped within the transport hub, giving it a certain character and potential in urbanization. This means that the designers need to translate it into a real place, with the characteristics of a modern city.

Xizhimen transport hub in Beijing, covering 4.52 hectares, serves the function of allowing people to transfer between rail transit and transit by bus, car and bicycle. It also has one office building and two residential buildings. The total floor area is 280,000 sq m, of which 80,000 sq m is for shopping, and it uses a most simple design, offering a convenient transport system to organize all aspects of horizontal and vertical flow. Sunlight is introduced to the underground space, creating a more comfortable environment when below ground. It provides the opportunity of working, shopping and recreation in one complex building, which will also help to reduce the demand for travel.

BICYCLES PLAY AN IMPORTANT ROLE IN URBAN TRANSPORT, BUT ARE VERY CONTROVERSIAL

Highly compacted and mixed land use generates shorter journeys. In many cities in China, bicycle travel accounts for a high proportion of mechanized travel, as high as 50–80%. There is no waiting time; it is very flexible and reliable, and bicycles clearly have the advantage over buses within a range of 4–6 km travel distances. The state code for urban road design prescribes the use of dedicated bicycle lanes on the major roads, so that our cities have the largest, best integrated bicycle system in the world. It is a very good energy-efficient solution. Many of China's cities can be listed at a very high level with regard to their energy consumption in urban transport. Nowadays, facing the shortage of petrol energy in the world, it is increasingly important to keep the use of the bicycle.

However, policies on the development of bicycle transport are among the most controversial. In the early 1980s, western countries advocated the control of private vehicles and encouraged public transport. At that time, China did not have so many cars as nowadays. The only private vehicle was the bicycle. So, many cities borrowed the policies for car control and applied them to bicycle control, trying to force people to abandon the bicycle to the slow speed bus.

In the 1990s, it was argued that the bicycle was an outdated mode of transport; traffic engineers considered that it was the major reason why traffic congestion was caused. So that its use was seen as the main obstacle stopping traffic improvements and hence the use of the bicycle had to be reduced, leaving more space for motorized vehicles. Nevertheless for a long time, the bicycle has fitted in well with the urban spatial structure and life rhythm.

In Shanghai, the ratio of bicycle over bus was 30:70 in the early 1980s, but by the early 1990s this ratio had reversed to 75:25.[5] There was a big drop in the numbers of bus passengers, while bicycle ownership was as high as 9.37 million. Furthermore, with the expansion of the city, travel distances became longer, and people began to choose the electric bicycle. In 2004, Shanghai had 835,000 electric bicycles, which is 40 times higher than the number in 2001.

Nowadays, the bicycle is still very widely used. Restrictions on bicycle use are clearly against the objectives of sustainable development, and are not conducive to social justice or the establishment of coexistence in a diversified urban society. It is worthy of mention that, after the failure in the strategy to "accelerate" the speed of motorized vehicles for several years, the Beijing Urban Planning Commission now intend to develop specialized "bicycle transportation planning" to encourage bicycle travel, even though there was in one year (2003) a big decrease of 2.9 million in the ownership of non-motorized vehicles.

INNOVATION FOR URBAN MOBILITY IMPROVEMENT

In the past 20 years, China has experienced tremendous achievements in improving urban mobility through transport infrastructure construction. However, due to the increase in travel demand and the greater diversity of urban life, the complexity of the issues arising from urban transport is far greater than expected. In 2003, the speed on some of the main roads in Beijing, in peak travel time, dropped to 12 km/hr, and even went as low as less than 7 km/hr. In Shanghai city center, "three vertical and three horizontal" main thoroughfares and key roads in peak hours has still only attained an average speed of 10–18 km/hr. Urban mobility seems to be returning to the era of the horse-and-carriage.

The bias that exists in the understanding of urban mobility and traffic flow will make it very difficult to bring about diversified and dynamic urban development.

The successful experiences have demonstrated that urban mobility development must adapt to a city's socioeconomic development and its cultural, historical and environmental characteristics. We really should urge more innovative thinking in urban mobility.

Chinese scholars have noted that the characteristics of the urban transport development environment mean that Chinese cities are facing the pressures of rapid urbanization, industrialization and motorization simultaneously, while also having to address the constraints in oil resources, land resources and institutional segregation. Some special aspects will be mentioned in the following, with the intention of gaining attention in the study of urban mobility.

1. The integration of inter- and intra-urban area transport

Previously, according to the planning economy, the function of an urban area should be quite self-contained, so that there was a clear territory boundary of transport service by the intra- or intercity transport provider, which belonged to different ministry authorities with different objectives and functions. Nowadays the megacities in China are experiencing a transformation in spatial structure from an isolated, territory-based city to a more network-based metropolitan area.

In the Yangtze River Delta, the Pearl River Delta and Beijing–Tianjin–Tangshan, etc., the spatial structure of an interlocking megalopolis is emerging. The linkage of socioeconomic activities is far beyond the boundary of the limited conventionally designated central city area. The integrated development of a region needs to be supported by a more coordinated metropolitan area-wide transport strategy. The conventional territory boundary within the central city for urban transport planning should be extended to have more coherence with intra-/intercity transport.

The Pearl River Delta region rail transport network planning and the Yangtze River Delta intercity rail transport improvement strategy reflect this intention of the cohesion of the transport system. The conventional function definition of the intra-/intercity transport has been an obstacle to mobility improvement in China.

2. Building a multimodal transport system, reducing on car travel

Accompanied by rapid economic development, the growth in travel demand is tremendous, and will require a fundamental improvement in urban mobility following the principle of sustainable development. The total trips made in Beijing were 5.364 billion in 1990 and rose to 6.928 billion in 2000 (the previous urban transport plan only expected 6.3 billion in 2000 and 6.8 billion by the year 2010). There was a 100% increase in trips from 1984 to 2003 in Guangzhou.[6] The travel demand has far exceeded expectations. How to manage and meet the demand has demonstrated tremendous differences between cities. From 1997 to 2000, ownership of civil vehicles in Beijing increased from 271,000 to 1.03 million (43% higher than predicted) and climbed to 2.35 million in 2005. The policy of facilitating the car reduced the mobility of the city.

The vehicle license auction policy is quite controversial in China, but it is combined with other measures including high parking fees, constrained car ownership and car use, maintenance of a reasonable level of urban mobility and promotion of the coexistence of multi-modal transport systems of bicycle, metro, bus and taxi, all of which play their part. Under the environment of a well-established multi-modal transport system, car travel is only one of the options, and it can be substituted by a more sustainable mode of transport. We could also ensure the efficiency and competitiveness of a city, with very limited car travel under an innovative sustainable urban transport strategy.

3. Challenges of the aging society and a more diversified city life

Until now, the urban transport system has been built mainly for working age people and commuters. Currently, China's population over the age of 60 has reached 130 million, meaning China has the world's largest elderly population. By the middle of the twenty-first century, China's elderly population will be over 400 million, roughly a quarter of the total population. At the same time as considering urban expansion, we should consider their need for travel in order to go shopping, to visit friends, for health care and for accompanying their grandchildren.

With the improvement of people's living standards, there will be more leisure and recreational activities; there will be an increasing demand for people to travel with flexibility, comfort and individuality. Every weekend night, after shop closing times, there are always long queues on the streets waiting for taxis – it takes a very long time for people to disperse.

This reflects the mismatch between the transport systems designed for mass commuting – with fixed routes, fixed schedules and fixed services – and more diversified demands. Another issue lies in weekend travel to the suburbs for recreation. The Shanghai tourist distribution center provides high-quality bus services to the suburb recreation destinations, which also constrains the desire to have a car.

4. Priority to improve the pedestrian environment in city centers

Our cities are fragmented and split apart by the ever more complicated flow of traffic. The main criterion for the design of urban streets is traffic volume, with roads graded according to the volume of cars passing through, but not for the convenience of pedestrians. The cities are becoming engineering structures in which only motorized vehicles can move freely. A survey shows that 44% of children on the way to school had encountered very dangerous traffic situations in Shanghai and Guangzhou, and 60% of them think that it is difficult to cross the road. A more civilized city is not one with a huge road traffic facility, but one that is a safe and comfortable place for the people.

5. Social capital resource and soft measures

The construction of infrastructure is important in order to improve urban mobility, and there is a real shortage of it. But we should also use social capital resources and some soft measures, for example, car-pooling with colleagues can easily be accepted as the same quality of travel experience of driving alone. With the provision of delivery services in shopping malls, we may choose to return home by bus or metro instead of taxi or car. In residential areas, people can share the parking spaces of office buildings in the vicinity, in order to save land space. Using a bicycle connecting service (renting or parking) could extend the metro service coverage area to ten times larger than for walking only. Except for traffic, conventional urban transport planning lacks studies on these issues.

6. Improve the design quality of urban flow space

In the design of urban flow space, such as motorways, multi-level interchanges, terminals and parking garages, we should give priority to the efficiency of the flow of traffic and flow of people, but quite often the engineering-driven concept dominates the whole process, which will make buildings and structures lack a sense of "place". There are so many people gathered together in stations and terminals, we should provide them with good quality space in these buildings, and this will also make public transport more attractive. Hence, at the design stage the full exchange of such ideas between engineers, architects and urban planners should be encouraged.

CONCLUSION

China has made great achievements in urban transport and mobility improvement, though not without experiencing some notable failures. It is important to review these valuable experiences in order to build a more sensible urban transport strategy for the future. The land-use control policy and low level of motorization in many cities opens up the possibility for us to implement a more innovative urban mobility strategy.

REFERENCES

1. Pan Haixiao (Ed.), *Architecture on the Move*, China Construction Press, 2004.
2. Pan Haixiao, *Shanghai 2010 Expo Transport Planning Concept Study*, Urban Planning Forums, 2005, 1 (China).
3. Shanghai Urban Comprehensive Transport Institute, *Shanghai Urban Comprehensive Transport Plan*, Shanghai Urban Comprehensive Transport Institute, 1992, 12.
4. Ministry of Construction China, *China Urban Construction Statistic*, Ministry of Construction China, 2003.
5. Xu Xuncun, The Study on Urban Transport Structure, in *Urban Transport Mode and Multi-Modal Interchange*, edited by Haixiao Pan and Jean-François Doulet, Tongji University Press, 2003.
6. Guangzhou Transport Plan Institute, *2003 Guangzhou Transport Annual Report*. Guangzhou Transport Plan Institute, 2003.

Chapter 15

The potential of casualty prevention in road traffic

Matthijs J. Koornstra

CONTENTS

ROAD INFRASTRUCTURE SAFETY IMPROVEMENTS

Road users' behaviour is not less or more responsible on different road types. Yet road risks differ markedly, dependent on the traffic complexity created by the road type and the allowed road modes, as shown for Germany and the Netherlands in Table 15.1. This table shows that it is not speed that kills primarily, because the lowest fatality risks are in calming areas (30 km/h limit) and on motorways (100/120 km/h limit in the Netherlands or no limit in Germany). On these road types, the speed variations between road users are proportionally smaller than on other road types. The injury risk is highest on roads where relatively high differences in speed and direction are present in combination with moderate speed limits (50 or 70 km/h limit, mixed slow and fast traffic, level crossings, opposing traffic). The fatality risk is also highest where the speed limits are inappropriately high (e.g. the 80 or 100 km/h limits on roads with mixed traffic, level crossings, and opposing traffic without mid-barriers).

Clearly, the infrastructure design and traffic rules of roads determine the traffic complexity for its road users and, thereby, the risk differences between road types. These differences are mainly explained by the effects of different average impact speeds in crashes, especially in crashes with pedestrians or cyclists, and by the effects of speed differences on crash frequency. Figure 15.1 explains why slow traffic should never be mixed with relatively fast motor traffic and should never cross roads where motor vehicles are driving faster than 30 km/h.

Table 15.1 Risks per vehicle kilometres on road types in the Netherlands, 1994, and Germany, 1993

Road type	Speed limit		Mix	Crossing/opposing traffic	Injury rate 10^6 km		Fatality rate 10^8 km	
	Netherlands	Germany	Fast/slow		Netherlands	Germany	Netherlands	Germany
Calming area	30	–	Yes	Yes	0.20	–	0.3	–
Residential roads	50	50	Yes	Yes	0.75	1.75	1.0	1.7
Urban arteries	50/70	50/>50	Yes/no	Yes	1.33	1.37	2.3	2.1
Rural roads	80	100	Yes	Yes	0.64	0.44	3.6	2.8
Rural arteries	80	100	No	Yes	0.30		1.8	
Rural motor roads	100	100	No	Yes/no	0.11	0.26	1.0	0.8
Motorways	100/120	No	No	No	0.07	0.15	0.4	0.5

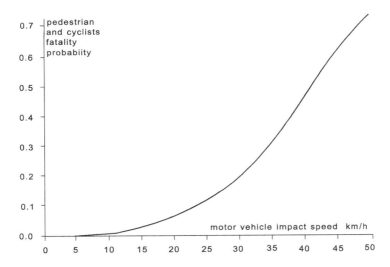

Figure 15.1 Pedestrian/cyclist fatality probability as function of impact speed.

Figure 15.2 demonstrates that speed is quadratically related to crash involvement on main rural roads and expressways, although old results for the United States, summarised by Warren (1982), differ for speeds below average speed from recent results in the UK (Taylor et al., 2000, 2002).

The probability of crashes with a fatality outcome depends by physical law in a quadratic way on the collision speed itself. The crash probability on a road type also depends mainly on its average speed because of the

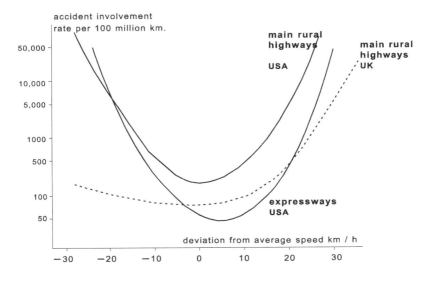

Figure 15.2 Crash involvement rate as a function of motor vehicle speed differences.

dependence of crash avoidance (reaction time, braking distance, and steering control) on the speed driven and also, of course, on the probability of a traffic conflict, which latter probability depends mainly on speed differences that cause overtaking manoeuvres, rear-end conflicts, and increased traffic complexity for other road users. Since speed differences generally also increase proportionally with the average speed a proportional change in average speed relates by a power of four to a proportional change in fatalities (e.g., a speed reduction of 7% results on average in a fatality reduction factor of $(0.93^2)^2 = 0.75$ or 25%), which is illustrated by Figure 15.3, that fits the observed road safety effects from speed reductions after lowered speed limits or intensified speed limit enforcement on several road types (from freeways to residential roads) in many countries. Since this relationship is almost fully determined by physical laws, it is astonishing that speed limits on motorways are still lacking or rather high in some countries and that many nations still have main rural highways with a limit of 90 or 100 km/h, which is too high for their mixture of allowed traffic.

Therefore, road safety is markedly enhanced if

- No mixing of fast and slow traffic is allowed on roads and crossings with car speeds that are higher than 30 km/h. Thus, mixed traffic should only be allowed on roads and crossings in 30 km/h areas, and only on their intersections with 50 km/h roads if constructed as single

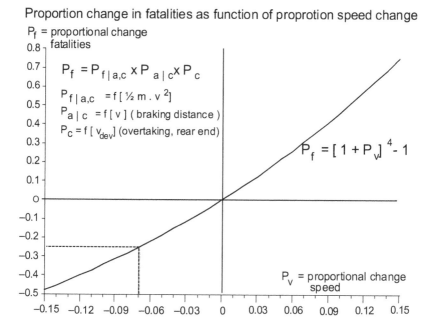

Proportion change in fatalities as function of proprotion speed change

P_f = proportional change fatalities

$$P_f = P_{f\,|\,a,c} \times P_{a\,|\,c} \times P_c$$

$$P_{f\,|\,a,c} = f\,[\,\tfrac{1}{2}\,m \cdot v^2\,]$$

$$P_{a\,|\,c} = f\,[\,v\,]\ (\text{braking distance})$$

$$P_c = f\,[\,v_{dev}\,]\ (\text{overtaking, rear end})$$

$$P_f = [\,1 + P_v\,]^4 - 1$$

P_v = proportional change speed

Figure 15.3 Dependence between proportional speed and fatality changes.

lane roundabouts of a 30 km/h design that have separated cycle paths and footpaths. The 50 km/h roads in built-up areas need to have separated cycle paths and footpaths, while intersections of 50 km/h roads should also be constructed as single lane roundabouts of a 30 km/h design that have separated cycle paths and footpaths.

- Rural roads up to 80 km/h and main residential arteries of 70 km/h have separated cycle paths, mid-barriers, graded level crossings for cyclists and pedestrians, and single or semi-dual lane roundabouts for motor vehicles that locally reduce their speeds to 50 km/h (no crossings).
- Motor roads with speed limits of 80 km/h or higher have mid-barriers, graded level crossings, and sufficiently long exit and entry lanes, while freeways with limits above 80 km/h should also have shoulders and additional lay-bys within 2 km for emergency stops.

If such a redesigned road infrastructure would be accomplished then the result would be that pedestrian and cyclist fatalities would reduce by at least 90%, while motorised road user fatalities on urban and rural roads would reduce by about 75%, on motor roads by 60%, and on freeways by 30%. In total, about 80% of fatalities could be saved by such a road infrastructure redesign. The full reconstruction is costly and will take a longer period than two decades, but many redesign measures that are most effective and less costly can be taken within a decade and then could likely save at least 40% of fatalities.

Figure 15.4 presents a fully redesigned road structure schematically for a city of 75,000 citizens, wherein we have not pictured the streets of the 30 km/h areas that are surrounded by 50 km/h roads with 30 km/h roundabouts. The road design within the central square in Figure 15.4 would apply to a village of about 10,000 inhabitants. Notice that, here, cyclists and pedestrians never meet cars with speeds above 30 km/h in the whole city or village, while crashes with vulnerable road users in 30 km/h areas are hardly fatal. Nowadays such a reconstructive road design for the application of the infrastructural improvement principles discussed above is to a large extent the kernel of the sustainable safe road transport policy of the Netherlands, as well as for the Swedish and Swiss Vision Zero policies on road safety, but it should also be the guiding principle for future road infrastructure development and reconstructive road rehabilitation, especially in newly motorising or medium motorised countries, where the majority of road fatalities and serious injuries concern vulnerable road users.

SAFETY IMPROVEMENTS FOR ROAD USER BEHAVIOUR

Not wearing seat belts by car occupants, no helmet use by (motorised) two-wheelers, drinking and driving, and speed limit violations are four

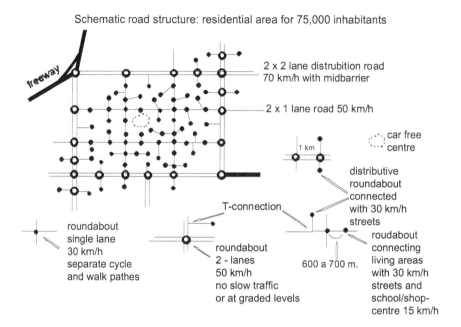

Figure 15.4 Schematic road infrastructure redesign for sustainable, safe traffic in a city.

behaviour types that also contribute to a high fatality risk. The effectiveness of seat belt wearing seems misunderstood, because the frequency of seat belt use generally decreases with the speed limit of the road, whereas in fact seat belt effectiveness is higher when the crash impact speed is lower, as Figure 15.5 (Koornstra, 1994) shows.

The curves in the figure derive from US data for 1980 to 1990, but nowadays seat belt injury prevention is highest on 50 km/h roads and their fatality prevention is highest on roads with moderately higher speeds, because in modern cars that satisfy New Car Assessment Programme (NCAP) norms, seat belt wearing generally prevents a fatality outcome for crashes with a collision speed below 65 km/h. Therefore, if higher speeds than 70 km/h are allowed, then the chance of crashes must be reduced by mid-barriers, graded level crossings, and by intensified speed limit enforcement that reduces speed and speed variation.

The effectiveness of the intensified enforcement of illegal road user behaviour is shown in Figure 15.6, taken from the SUNflower Report (Koornstra et al., 2002). In this figure the empirical data for enforcement intensity on drinking and driving and on seat belt wearing in Sweden, the UK, and the Netherlands are plotted along the theoretical function for the reverse-S-shaped relationship between enforcement intensity (annual number of controls per licence holder) and level of violations (different for different types of violations). Without police control the level of not wearing a seat belt

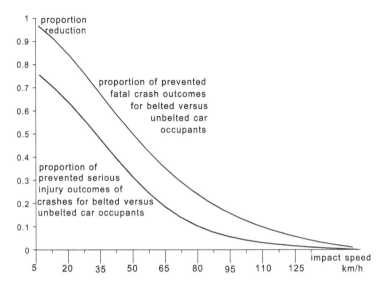

Figure 15.5 Seat belt effectiveness as a function of collision impact speed.

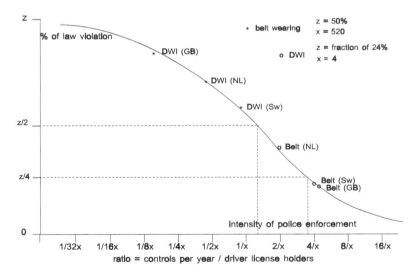

Figure 15.6 Police enforcement intensity and its effectiveness.

is about 50% (level $z = 50\%$ in Figure 15.6 for this type of violation), but already about 1 annual control per 65 licence holders (level $8/x$ with $x = 520$ for seat belt use controls) reduces the level of seat belt wearing violation to about 6%. Higher seat belt use enforcement intensity could save more than 30% of fatalities in the United States, because about 35% of car occupants are unbelted in the US, with the result that about 60% of all US road fatalities are unbelted car occupants.

It is well known that the fatal crash probability increases exponentially with the blood alcohol content (BAC) of the driver. Without a noticeable police control the violation level of the driving and drinking law is generally about 24% (level z for driving while intoxicated (DWI) in Figure 15.6) for drivers above 0.1% BAC on weekend nights, which is then generally associated with about 40% of the national road fatalities. Most motorised countries have 0.05% or 0.08% BAC laws, but if the legal BAC level is 0.02% (as it is in Sweden), and if its police enforcement intensity by random breath testing would be as high as one annual control per licence holder (level $4/x$ with $x = 4$ for DWI), then probably about 30% road fatalities would be saved in general, where one annual control per four licence holders (the DWI-control level for Sweden) already reduces the fatal crashes from drinking and driving to below 15%.

The average speeds driven tend to be the same as the speed limit on roads where speed limit enforcement is hardly applied, whereby speed limits without police control are generally violated by about 50% of the drivers ($z = 50\%$ for speeding in Figure 15.6). Tentative Dutch results on the effectiveness of intensified speed limit enforcement indicate that speed enforcement intensity must be increased to more than four controls per year per licence holder in order to reduce speed limit violations to below 10%. If this could be achieved on urban roads and arteries with respective limits of 50 and 70 km/h and on main rural roads with an appropriate speed limit (preferably 80 km/h, except on rural motor roads), and if an appropriate speed limit on motorways could also be installed (preferably 100 km/h around big cities and 120 km/h or 130 km/h elsewhere) and controlled, then the average speeds and speed differences would be reduced by more than 12% in motorised countries, whereby the total of the fatalities in these countries would then be reduced by about 40%.

As Figure 15.6 shows, if the existing enforcement level is low, then a four-fold intensified level (from $1/64x$ to $1/16x$ in Figure 15.6) will hardly show any effect. However, if the existing enforcement level has already reduced the violation level to 75% of its level without controls (control level $1/4x$ with violation level $3/4z$), then a fourfold intensified enforcement level (to $1/x$) will reduce the violation level by 33% (from $3/4z$ to $1/2z$). It explains why the effectiveness of intensified police enforcement has shown contradictory results, but intensified police enforcement can be very effective, provided that its enhanced intensity is sufficiently high. Moreover, targeted information campaigns that sustain specific enforcement types have shown to increase the effectiveness of enforcement (Koornstra and Christensen, 1990). General information campaigns and road safety awareness actions have shown no direct safety effects, but may be indirectly needed for public acceptance and for the political priority of road safety measures. In addition, road safety education has minor effects, because the risk reduction for new users of each road mode to the three- to fourfold lower risk levels of experienced road users requires several years of daily practice in traffic (driving 100,000 km for newly licenced drivers). Only changes to graduated driving licencing systems (for example, only driving when accompanied by an experienced driver in the first phase and initial restrictions, such as no alcohol, no night driving, no young passengers) has shown 10% to 40% risk reduction (Sensenck and Welan, 2003), but two years after the full licence the maximal effect is 20% less crash involvement of new drivers, which means nationally less than 4% casualty reduction.

Thus, road safety can also be markedly improved by intensified police enforcement of: (1) helmet wearing by motorised and non-motorised two-wheelers; (2) seat belt use by car drivers, including front and back seat passengers and appropriate child restraint use on the back seats; (3) drink driving laws; and (4) observation of speed limits. Intensified police enforcement – with respective fatality reductions up to 30% for seat belt and helmet use, 25% for drink driving observance, and 40% for speed limit observance (if appropriate limits are installed) – could together reduce the fatalities by $100*[1 - (1 - .30)*(1 - .25)*(1 - .40)] = 68\%$ in motorised and motorising countries. Thus, over 65% fatality reduction is achievable within less than a decade by intensified enforcement and more appropriate traffic safety laws. The state investment for the necessary higher enforcement level is not a problem, because it would be less than the increased state revenues from traffic fines.

VEHICLE SAFETY IMPROVEMENTS

Improved passive vehicle safety has probably saved about 20% of fatalities in the last two decades (Broughton, 2003; Koornstra et al., 2002) in the European Union (EU). Car renewal and improved passive vehicle safety could further save more than 20% of fatalities in the next two decades, if, in time, priority would be given to new effective vehicle safety regulations, such as an automatic ignition block if someone is not belted, soft-nose car construction for vulnerable road user protection, compatibility requirements for car and freight vehicles, under-run protection for trucks and lorries, etc. Also regulations for active vehicle safety that have been proved to be effective, if timely taken and if retrofitted, could save more than 30% of fatalities. Most effective, in the short run, are mainly vehicle devices that improve risk perception or keep drivers within legal limits, such as intelligent speed adaptors, automatic daytime running lights (DRL), collision avoidance assistance devices, driving data recorders with feedback information, etc. Intelligent speed adaptors save probably between 20% to 40% of fatalities (Várhely, 2003; Oei and Polak, 2002) and for driver monitoring by in-car data recorders the figure is probably over 20% (Wouters and Bos, 1997, 2000; Heinzman and Schade, 2003), while DRL saves dependence on the region's latitude saving approximately 8% to 20% casualties in multiple daytime crashes (also for crashes with cyclists and pedestrians). Figure 15.7 shows the result of a meta-analytic study on effects of DRL use, reported to the Directorate General for Energy and Transport (DG-TREN) of the EU (Koornstra et al., 1997).

With regard to intelligent vehicles (CEC, 2003), however, intelligent driving systems may only become effective in the long run. If such systems are based on communications between vehicles then they cannot be implemented in fail-safe way, because the majority of vehicles will not be equipped in the early years following their introduction. Only intelligent systems that also offer additional safety in the present traffic systems should be supported. In the last 20 years, there has been a massive amount of research undertaken on safety devices in vehicles based on information and communications technology (ICT) and on automatic or ICT-assisted driving, but as yet there are still no fail-safe or markedly safety-improving systems. Therefore, no marked safety gains are expected from such ICT-based driving systems before 2020. The further use of already regulated, passive safety devices, as well as other devices that are known and available for new passive and active vehicle safety regulations, can together save 50% of fatalities, but new regulations apply generally to new vehicles. Therefore, it is of the utmost importance that authorities take action as soon as possible on new vehicle safety regulations and, where possible, also on retrofit implementations (e.g., DRL) for available vehicle devices that have a proven

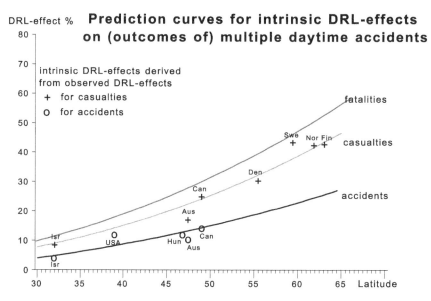

Figure 15.7 Effects of DRL changes from 0% to 100% as a function of average country latitude.

effectiveness, such as those discussed above. However, even if such new vehicle safety regulations are taken soon and active devices are retrofitted then they still cannot save more than 25% of fatalities within two decades.

THE FEASIBLE FATALITY REDUCTION

The discussed road reconstruction, intensified police enforcement, and vehicle safety could save about 40%, 65%, and 25% of fatalities, respectively, within two decades in motorised countries, which then would save $100*[1-(1-.40)*(1-.65)*(1-.25)] = 85\%$ of fatalities (correcting for fatalities already saved by other measures). Also Elvik (2003) derived independently a potential casualty reduction of about 80% by existing, but inadequately applied measures. In the long run, and if fully applied, the respective measures (with their separately maximal 80%, 70%, and 50% effect) could save nearly all fatalities, since it means together $100*[1-(1-.80)*(1-.70)*(1-.50)] = 97\%$ reduction of fatalities without the additional post-crash effects of optimised rescue and trauma care. However, the full implementation of the recommended road infrastructure reconstruction will take a very long time, because it must be financed by long-term investment plans that need to combine road maintenance and rehabilitation (that usually affect all roads within 40 years) with the recommended infrastructure reconstruction, in order to reduce the reconstruction costs and to be able to finance the total road investment plan. The fatality reduction from intensified police enforcement and adjusted laws is, in principle, achievable in a relatively short period, but the social and political climate must change before they can be realised, and the required level of enforcement intensity may realistically not be reached fully in the short term. The cumulative effect of vehicle safety measures will take also decades, partially due to the required full renewal of the national vehicle fleet. Thus, the maximum achievable fatality reduction can only be realised in the long term, but in view of the potential to achieve 97% fatality reduction it seems feasible to realise a zero-vision approach by use of a road safety policy that is based on long-term investment plans for a sustainable, safe reconstruction of the road infrastructure, multiple intensifications of police enforcement of optimised road safety laws, cumulative vehicle regulations for effective, passive and active safety measures, and post-crash rescue and trauma care measures.

The major obstacles for the zero-vision achievement of sustainable, safe road traffic are the lack of sufficient political priority for road safety, even in highly motorised countries (Elvik, 2003), the lack of awareness for the road safety problem, especially in not so highly motorised countries, and the general ignorance of the existence of effective road safety measures by authorities. These obstacles are the reasons why the proposed, effective road safety measures are infrequently implemented, despite the fact that they would save many lives, as well as reduce the level of seriously injured road victims, many of whom are lifelong completely or partially disabled, and also despite the fact that they would increase the welfare of the country, because the proposed road safety measures have a positive macroeconomic return on investment (Koornstra, 2002a; Elvik, 2003). Political priority for the solution of these road safety problems is largely absent, not only in relatively less motorised countries, but this is not caused by political ignorance or at the wish of inhabitants. Also road users are largely indifferent to road traffic risks and regard road accidents as random events – as inevitable consequences of the benefits of motorised road transport. On the one hand there is the positive utility of motorised and non-motorised road use, but on the other hand no one likes the risk of being involved in a road accident. However, the cake cannot be both saved and eaten. The ambivalence of the conflict between the understood aspects of road transport and the dislike of road risks becomes psychologically resolved by partial risk indifference. Moreover, differences in road traffic risks are difficult to perceive. Only road traffic situations with unusually high or low risks are consciously perceived, where extremely high risk situations are generally compensated for by safer behaviour, but regrettably the reverse often holds for extremely safe road situations. Also, road crashes are not disasters that kill and injure a mass of people in a single event, but road injuries or deaths occur separately and generally concern unknown, distant individuals. People become accustomed to an average risk level in more or less the same way (but much more slowly) as, for example, in darkness adaptation. Road users, therefore, are to a large extent indifferent to a range of risks around the average road traffic risk (Koornstra, 1990, 2007). This adaptation to experienced average risk and the ambivalent indifference for a risk range around average risk is what causes the ignorance of road safety problems by the majority of people and politicians, who both also believe, wrongly, that there are not really any effective measures to reduce the problem.

Nonetheless, the road safety problem is not caused by acts of god, nor is it out of the reach of human control. It is a man-made problem that can be reduced by human actions. Politicians ought to act rationally, and should not be guided by psychologically understandable, but wrong, reasons for disregarding the road safety problem. They need to become better informed about the human impact and economic costs of the road safety problem and about the urgency and potential of road casualty prevention. At the individual level, road traffic risks may be ignored and regarded distant from us, but nowadays road crashes mean that, worldwide, more than 1.2 million road users die and almost 7 million are seriously injured, while the national costs of road crashes range from 1.5% of the BNP in highly motorised and hardly motorised countries to 2.5% of the BNP in medium motorised countries (Koornstra, 2002a,b). Moreover, the world total of road fatalities is predicted to increase, if present trends continue (WBCSD, 2004). Preventative actions are not only the responsibility of politicians in parliament and transport ministers. Regional and local authorities, decentralised organisations, and all who are professionally involved in road transport (planning, management, construction, and the vehicle industry) as well as relevant other sectors (health care, legislation, police, and education), have to face the road safety problem rationally and take the necessary preventative actions. If cost-effective countermeasures did not exist, then there would be an excuse for the lack of preventative action, but effective road safety measures with excellent cost-benefit ratios are professionally well known in highly motorised countries. Therefore, it is also of utmost importance to transfer that professional knowledge to the authorities in highly motorised countries, if they are ignorant of them, and even more so to uninformed relevant bodies and professionally involved people in medium motorised and newly motorising countries, which all have relatively much higher road user risks.

REFERENCES

Broughton, J. (2003). The benefits of improved car secondary safety. *Accid Anal Prev.* 35: 527–535.

CEC. (2003). Information and communication technologies for safe and intelligent vehicles. COM (2003) 542 final. Commission of the European Communities, Brussels, Belgium.

Elvik, R. (2003). How would setting policy priorities according to cost-benefit analysis affect the provision of road safety? *Accid Anal Prev.* 35: 557–570.

Heinzman, H.-J. & Schade, P.D. (2003). *Moderne Verkehrssicherheitstechnologie: Fahrdatensspeicher und junge Fahrer.* Berichte der Bundesanstalt für Straßenwesen (BASt), Heft 148. BASt, Bergisch-Gladbach, Germany.

Koornstra, M.J. (1990). System theory and individual risk. pp. 21–45. In: Benjamin, T. (Ed.). *Driving Behaviour in a Social Context.* Paradigm, Caen, France.

Koornstra, M.J. (1994). The dependence of belt effectiveness on delta V. Research Note. SWOV, Leidschendam, Netherlands.

Koornstra, M.J. (2002a). The global estimation of the health and economic losses from road accidents, their prognosis for 2020, the prevention potential, and the health and economic prevention benefits. Unpublished research report for the WHO, World Health Organisation, Geneva, Switzerland.

Koornstra, M.J. (2002b). The urgency and potential for prevention of road traffic deaths and injuries. Unpublished report for the WHO, World Health Organisation, Geneva, Switzerland.

Koornstra, M.J. (2007). *Changing Choices: Psychological Relativity Theory.* Leiden University Press, Leiden, Netherlands.

Koornstra, M.J.; Bijleveld, F.D. & Hagenzieker, M.P. (1997). The safety effects of daytime running lights. Report R-97-36. SWOV, Leidschendam, Netherlands.

Koornstra, M.J. & Christensen, J. (Eds.) (1990). Enforcement and rewarding: Strategies and effects. Proceedings of the of OECD/EMCT Road Safety Symposium. SWOV, Leidschendam, Netherlands.

Koornstra, M.J.; Lynam, D.A.; Nilsson, G.; Noordzij, P.; Petterson, H.-E.; Wegman, F. & Wouters, P. (2002). SUNflower: A comparative study of the development of road safety in Sweden, the United Kingdom, and the Netherlands. TRL-VTI-SWOV report. SWOV, Leidschendam, Netherlands.

Oei, H.L. & Polak, P.H. (2002). Intelligent speed adaptation (ISA) and road safety. *IATSS Res*. 26: 45–51.

Sensenck, T. & Welan, M. (2003). Graduated driver licensing: Effectiveness of systems and individual components. Report no. 209. Monash Univ. Acc. Res. Centre, Victoria, Australia.

Taylor, M.C.; Baruya, A. & Kennedy, J.V. (2002). The relationship between speed and accidents on rural single-carriageway roads. Report 511, Transport Research Laboratory, Crowthorne, UK.

Taylor, M.C.; Lynam, D.A. & Baruya, A. (2000). The effects of drivers' speed on the frequency of road accidents. Report 421, Transport Research Laboratory, Crowthorne, UK.

Várhely, A. (2003). Dynamic speed adaptation in adverse conditions – A system proposal. *IATSS Res*. 26: 52–59.

Warren, D.L. (1982). Speed zoning and control. Chapter 17. In: *Synthesis of Safety Research Related to Traffic Control and Roadway Elements*. Federal Highway Administration, Washington D.C.

WBCSD. (2004). Mobility 2030: Meeting the challenges to sustainability. Full Report of the Sustainable Mobility Project. WBCSD, Geneva, Switzerlands. Available at: www.wbcsd.org/Programs/Cities-and-Mobility/Transforming-Mobility/SiMPlify/Resources/Mobility-2030-Meeting-the-challenges-to-sustainability-Full-Report-2004.

Wouters, P.I.J. & Bos, J.M. (1997). The impact of driver monitoring with vehicle data recorders on accident occurrence. Res. Report R-97-08. SWOV, Leidschendam, Netherlands.

Wouters, P.I.J. & Bos, J.M. (2000). Traffic accident reduction by monitoring driver behaviour with in-car data recorders. *Accid Anal Prev*. 32: 642–650.

Chapter 16

Health effects of transport

Carlos Dora

CONTENTS

INTRODUCTION

Transportation is a major public health issue worldwide. The modes of transport people use for everyday life, how long and how often we travel, all have major implications for the health of individuals and of the population (British Medical Association, 1997). Transport activities impact on health, both negatively and positively, and transport policies are a key determinant of the health of urban populations. To achieve the health potential from transport interventions, health has to be clearly included on the urban transport and land-use policy agendas.

Infrastructure works in the nineteenth century led to major public health gains in urban areas, through water, sewage and housing improvements.

The challenge at the beginning of the twentieth-first century is to plan and implement land-use patterns and transport systems that avoid traffic injuries, avoid air and noise pollution, promote safe physical activity and promote social inclusion.

Although the evidence for the health impacts of transport is vast, it has not yet been effectively used to mobilize action in support for sustainable transport policies. This paper summarizes the range of public health impacts from urban transport and proposes new mechanisms to harness this information, so it can contribute to the public health of urban populations and the quality of the city environment, and encourage the adoption of sustainable urban transport policies.

THE BURDEN OF DISEASE FROM TRANSPORT

Current transport policies are a key determinant of the global burden of disease (WHO, 2002); road traffic accidents cause 1.2 million deaths worldwide and urban air pollution is estimated to cause around 800,000 deaths in urban areas every year (65% of which are in Asia). Transport is also a root cause of physical inactivity, which causes 1.9 million deaths every year, noise pollution, climate change and psycho-social well-being. The evidence for these impacts is summarized below.

AIR POLLUTION

There is good evidence that transport-related air pollution is associated with short-term and long-term increases in ill health and mortality (Krzyzanowski et al., 2005), including cardiovascular diseases (ischaemic heart disease, cerebrovascular disease, hypertensive disease, inflammatory heart disease), respiratory diseases (chronic obstructive pulmonary disease [COPD], asthma), upper and lower respiratory infections and lung cancer (Cohen et al., 2004).

These health effects can be attributed to a mixture of pollutants. Particles with diameters of less than 10 or 2.5 microns in diameter are a good indicator of the air pollution mix that people are exposed to, and the smaller sizes can reach deeply into the lungs. Particles are emitted by road vehicles and are composed of carbon, heavy metals and carcinogens such as benzene. Diesel vehicles are usually responsible for a large share of particle emissions.

Increasing concentrations of particulate matter (PM) are associated with increases in health impact; the effect is stronger for PM 2.5, the particles coming from combustion engines, with a 3.4% increase in mortality for every 10 $\mu g/m^3$ of PM 2.5 and 0.6% increase for PM 10 (Schwartz et al., 2002).

No threshold could be identified below which health effects were not found, even for concentrations as low as 8 µg/m³ for PM 2.5 and 15 µg/m³ for PM 10. To facilitate air quality management, the World Health Organization (WHO) air quality guidelines set a target value of 10 µg/m³ of PM 2.5 and 20 µg/m³ of PM 10 annual average, clarifying the health benefits that can be obtained by reaching those levels (WHO, 2006).

Findings from studies of the health impacts of air pollution carried out outside Europe and the United States show around a 1% increase in short-term mortality for a 10 µg/m³ increase in PM 10. For example, cities with levels of 150 µg/m³ of PM 10, of which there several in the developing world, would have a 10% higher daily mortality than cities with 20 µg/m³ levels of PM 10.

Small particles can get indoors freely and can travel long distances across national boundaries, so neither the indoor environment nor distance from roads offer much protection to PM.

Other measures of exposure to air pollutants such as residence near busy roads, or self-reported traffic intensity at a residence, were associated with severe health outcomes (Roemer and van Wijnen, 2001; Hoek et al., 2002).

A large proportion of air pollution emissions is attributed to transport, especially in urban areas with large volumes of road traffic (EEA, 2002). An even higher proportion of exposure to air pollution may be due to transport as air pollutants concentrate in the immediate areas around (up to 250 m away from) urban highways, and in street canyons (Buringh et al., 2002). In addition, levels of air pollutants in vehicles (underground trains, cars and, to a lesser extent, buses) are up to a few-fold higher than around cyclists or pedestrians on the same street or in the same area (Sanderson et al., 2005). Faster breathing rates and a longer time of travel by bicycle in some European cities meant that there was no net difference in exposure to air pollutants between motor vehicle and bicycle users (WHO Regional Office for Europe, 2006). Journey time exposures contribute disproportionately to the total exposure, as people may spend hours a day commuting.

The health effects of air pollution observed in epidemiological studies seem to be greater in lower socioeconomic groups (Pope et al., 2002). One of the reasons for this phenomenon may be cumulative factors increasing the chances for personal exposures and/or greater personal susceptibility in lower socioeconomic groups (O'Neill et al., 2003).

Reductions in air pollution have been quickly followed by reductions in mortality, as observed in Dublin (Clancy et al., 2002) and Hong Kong (Hedley et al., 2002). These observations are consistent with time series studies (showing that daily variations in the health status of the population follow changes in air pollution) as well as with a study in California indicating improvement in lung function development in children moving to less polluted areas, and worsening lung function development among children moving to more polluted areas (Avol et al., 2001).

CLIMATE CHANGE

The transport sector is the fastest growing contributor to greenhouse gases (GHG). In 1990 it produced 13% of global GHG emissions and the total contribution is expected to more than double by 2020 (IPCC, 2001a). The Intergovernmental Panel on Climate Change (IPCC) considers that transport is also a sector where major opportunities exist for reductions in climate change gases, as substitutes and alternatives exist, and there is a wide range of co-benefits from transport policies that reduce GHG, by the direct impact on air pollutants, injuries and physical activity.

The burning of fossil fuels over the last 50 years has produced sufficient quantities of GHG to affect the global climate. The global average temperature has increased by around 0.6°C over the last century and is expected to go up by between 1.4 and 5.6°C during this century (IPCC, 2001b).

Climate change is an emerging threat to global public health. The impacts of climate change on health are mostly negative and highly inequitable, as they are expected to hit the poorest communities most severely, yet they have contributed significantly less to climate change (McMichael and Githeko, 2001). Health impacts from climate change are expected through extreme weather events, especially heat waves, floods and drought. Urban populations are especially at risk of exposure to heat waves in view of the heat-island effect of cities (Idso, Idso and Balling, 2001), where temperatures in the city are up to 5–11°C higher than in the surrounding rural areas (Aniello et al., 1995). Urban sprawl exacerbates that effect (Frumkin, 2002; Zhou et al., 2004), and cities in tropical developing countries do not show the compensation seen in wealthier cities from lower winter mortality and adaptation measures (Hajat et al., 2005). Changes in rainfall from climate change affect the supply of fresh water and increase the risk of water-borne diseases; and, together with higher temperatures, this change affects food production, increasing the risks of malnutrition. Vector-borne diseases may affect regions without population immunity, in view of changes in the length of transmission and altered geographical range. A clear example is dengue, the most important viral vector-borne disease globally. Dengue transmission has increased dramatically, especially in urban areas, in view of rapid, unplanned urbanization, expansion on the breeding sites for the Aedes mosquitoes and high human population density, which offers the supply of a large number of susceptible individuals (Gubler and Meltzer, 1999). Coastal area cities will also be affected by sea-level rise (McMichael et al., 2004).

City populations are therefore especially vulnerable to climate change health impacts. They are also partly responsible for climate change, and cities in developing countries are increasingly contributors of climate change gases. On the other hand, cities can reduce vulnerability to climate change health impacts, and they can also lower production of climate change gases through land use and sustainable transport measures that will have a number of co-benefits for public health and quality of life (through the abatement of traffic injuries, ambient air pollution, noise, severance and congestion). These connections are key to identifying the right policies for sustainable transport and public health.

TRAFFIC INJURIES

Almost 90% of the 1.2 million deaths and 20–50 million injuries due to road traffic every year occur in low- and middle-income countries (Peden et al., 2004). Around half of those deaths are among adults aged between 15 and 44 years. Road traffic injury (RTA) is already the ninth leading cause of death worldwide and is expected to become the third cause of death worldwide soon.

Pedestrians, cyclists and motorized two- and three-wheeler riders are the road users most vulnerable to traffic injury, and have the highest risk of death per km travelled of all transport modes (Peden et al., 2004). Eighty-five percent of traffic injuries occur in developing countries, and fatality rates by road traffic incidents (RTIs) in children in low- and middle-income countries is five times higher than in high-income countries (Nantulya and Reich, 2002). People aged over 60 years have the highest death rates per 100,000 population compared with other age groups in low- and middle-income countries (Peden, McGee and Sharma, 2002).

The bulk of road users in middle- and low-income countries will continue to be pedestrians, cyclists, motorized two- and three-wheeler riders and public transport users. It is essential that transport policies give priority to protect the safety of these user groups (O'Neill and Mohan, 2002) in order to promote public health and ensure health equity. These policies often focus on the private car user instead.

The World Report on Traffic Injury Prevention reviews effective policies and measures to prevent RTAs. These include land-use and transport policies such as: the promotion of high-density and mixed-use developments where work, residence and entertainment are all nearby, reducing the need for transport (Mohan and Tiwari, 2000; Litman, 2003); providing shorter and safer routes for vulnerable road users; and separating motorized and non-motorized traffic and discouraging private cars from entering city centres or encroaching into pedestrian space. Encouraging the use of safer modes of travel such as public transport, as well as cycling and walking, which pose less risk to others than do motor vehicles, are also effective, but in developing countries in particular the quality and safety of public transport services need to be assured, including through the use of regulation and enforcement. Enforcing speed and alcohol limits are among the most effective preventative measures, as is the enforcement of child restraints, safety belts and helmets (among users of two-wheelers).

PHYSICAL ACTIVITY

Developing and developed countries are facing an epidemic of non-communicable diseases (NCDs) (WHO, 2005b). Lack of physical activity and obesity are among the main root causes of NCDs, and a widespread increase in physical activity is one of the key recommendations to reverse that trend.

Cycling and walking for daily activities can bring major health benefits (Andersen et al., 2000; Hou et al., 2004) – half an hour a day can half the risk of developing heart disease, equivalent to the effect of not smoking. Even if spread over two or three shorter episodes, this amount of physical activity can half the risk of developing adult diabetes, of becoming obese or of developing certain cancers, as well as reduce blood pressure, reduce the risk of osteoporosis and improve functional capacity (US Surgeon General Report, 1996). These improvements can be expected for the 60% of the world population who do very little physical activity (WHO, 2005).

A large proportion of trips in urban areas are short and provide the opportunity for daily exercise required for health benefits, 15 minutes is on average the time needed for a 2 km walk or a 3–5 km cycling trip.

However, the enormous saving on the health care costs being provided by the high levels of cycling and walking for daily activities in many parts of the world are not recognized. There is now a trend in many cities in emerging economies to adopt transport policies that discourage cycling and walking, but these policies are likely to create a very substantial health burden in the future, as well as a large bill, considering the high costs of treating the range of chronic diseases caused by physical inactivity.

The risk of accident is an important deterrent to cycling. To obtain and maintain the levels of cycling and walking needed to protect public health, it is essential to guarantee safe cycling and walking. Under road safety conditions such as those found in the United Kingdom since the early 1990s, the health benefits from cycling were estimated to be 20-fold greater than the health risks associated with cycling (BMA, 1992; Rutter, 2005).

HEALTH AND THE BUILT ENVIRONMENT

The built environment has a key role in the promotion of physical activity and on public health, as demonstrated in recent research. Urban sprawl in the United States has been associated with obesity, lack of physical activity and hypertension (Ewing et al., 2003), but the observational study methods used do not permit the evaluation of whether there is a causal link. Intervention studies, on the other hand, are a much more robust method for demonstrating epidemiological links. Some examples of this stronger type of evidence exist for developed countries. For example, interventions to promote cycling showed a remarkable saving on health care costs (4.5 million euros) and many health benefits including a 20% reduction in road traffic injuries and a 20% reduction in mortality, from all causes, among 15–49-year-olds (Odense Municipality, 2004). Infrastructure investment to improve cycling and walking showed positive cost-benefit ratios of 4, 14 and 3 in three different Norwegian cities, (Sælensminde, 2002). The congestion charge instituted in central London led to a reduction in risks to health from air pollution and traffic injuries. An economic analysis of interventions to promote integrated non-motorized and public transport in Bogotá, Colombia, Morogoro, Tanzania and Delhi, show benefits were 5 to 20 times greater than the costs (Ice, 2000).

Overall there is a shortage of studies (with the exceptions of the studies in northern Europe described above) documenting the range of health impacts that follow transport and land-use interventions. At the same time, there are a number of significant interventions in urban transport and land use across the globe that could be used to investigate the actual health gains from transport. This gap is an obstacle to advancing urban public health and sustainable transport agendas.

CURRENT CHALLENGES FOR SUSTAINABLE TRANSPORT AND HEALTH GAIN IN URBAN AREAS

A large share of the risks to health from transport occur in urban areas. Urban dwellers are particularly exposed to those risks, as humans and vehicles share the same restricted spaces. Half the world's population already lives in cities. Over the next 30 years the population of the world's cities is expected to double to over 5 billion people, and almost all of this growth is taking place in developing countries. Health risks from transport thus constitute a public health problem of the very first order.

There is an unprecedented growth in the number of motor vehicles and traffic volumes in urban areas worldwide. Vehicle kilometres travelled and vehicle stocks are increasing more rapidly in non-OECD countries than in the OECD area (76% vs 33%, and 70% vs 42% between the years 1995 and 2010). Most of this increase is in private car use (OECD, 2001).

Transport policies in cities worldwide are designed with the private car in mind. Yet the vast majority of the world population will not have a car for the foreseeable future; they will suffer from traffic congestion and the health risks from pollution, traffic accidents and barriers to cycling and walking, but they will not benefit from private mobility.

Cities in the developing world often have a high proportion of daily trips by foot or bicycle – in Chinese cities around 60% in the mid-1990s and 40% in cities in Africa (Kenworthy and Laube, 2002). Urban transport policies discouraging non-motorized vehicles (NMVs) and privileging the car therefore create health inequalities, especially and increasingly in developing countries. The poor tend to suffer most from the risks of transport, as they are the group most reliant on cycling and walking for transport, live and work on the streets, travel longer distances to and from work, and use more hazardous vehicles, such as two-wheeler mopeds (Mohan and Tiwari, 2000).

Unhindered motorization shapes cities and communities (urban sprawl), increases travel distances, in turn engendering health risks (physical inactivity, longer exposure to pollutants and risk of injury), and triggers further urban sprawl and motorization. Busy streets have been linked to the loss of community cohesion and the limiting of children's development. Children are especially vulnerable to environmental pollution and injuries, and traffic noise can limit their learning ability.

In addition, urban transport policies have overlooked the health and security function provided by street vendors, and pedestrians in general, as streets with people using them are safer from violence and traffic injuries. Small street businesses provide livelihoods and, through that, contribute to the health of individuals and their families. Street vendors and others who work in traffic, such as police officers, suffer more from air pollution disorders and are at higher risk from traffic injuries.

Urban transport solutions that promote health

The experience in a number of cities has shown that better transport systems can be developed in relatively short time periods and address a wide range of issues, from congestion to social inclusion of different groups in the population.

Some of the key ingredients include:

a. Transport demand management – city planning that emphasizes the proximity between homes and work/markets/leisure/study places, reducing the need to travel and exposure to transport health risks.
b. Priority for transport modes with the least risks to health per unit of travel: low pollution, low injury risk and enhanced physical activity. Priority in the urban space should be allocated to public transport, cycling and walking, with connectivity between these modes assured.
c. Protection of vulnerable road users – pedestrians and cyclists should be assured dedicated space and protection in road design. Public transport and urban spaces should be accessible to all levels of physical ability.
d. Active community environments – space supportive of social interaction and physical activity,
e. Adoption of clean and safe technologies: improvement of vehicle maintenance standards, testing and surveillance of polluting vehicles; enforcement of speed and alcohol-while-driving limits and traffic regulations; and adoption of individual safety measures, such as helmets and seat belts.
f. Internalization of all the health-related external costs into transport decision-making.

WHY ARE TRANSPORT AND HEALTH NOT AT THE TOP OF THE POLITICAL AGENDA OF MANY CITIES?

A few of the problems in making this case are listed below.

Experience highlights the limitations of narrowly focused interventions in delivering the expected health benefits, as several health risks from transport are closely related. For example, the impact of technical solutions to reduce air pollution based purely on technology improvements may be quickly cancelled out by the added growth in the number of motor vehicles and increased travel distances (OECD, 2001). The change in health risks from adopting liquefied petroleum gas (LPG) in buses for a city, for example, involves substitution for other, more polluting means of transport such as two-stroke engine two-wheelers in view of the resulting bus price increases (Tiwari, 2003) .

Environmental assessments of transport policies often claim to lead to health improvements, but these assessments systematically overlook the health benefits of physical activity through transport, one of the main protection factors for the largest causes of death.

There is so far a major gap in understanding the contribution of cycling and walking for the transport system and for public health, leading to interventions to curb cycling and pedestrians from the streets on the basis that rapid transit is the only solution. The role of street vendors, and of non-motorized street users is also largely overlooked by system policies.

There is a focus on the technical solutions, but analyses of the decision-making process and of what promotes change are often lacking.

There is above all a large gap in the measurement of health impacts from transport interventions – often only one or two health impacts are considered, and the bulk of the health burden and benefits are ignored.

HOW TO MAKE SUSTAINABLE TRANSPORT SOLUTIONS A PRIORITY IS AS IMPORTANT AS HAVING THE TECHNICAL KNOWLEDGE ABOUT WHAT NEEDS TO BE DONE

By and large, the technical answers to transport problems are already known by experts, and in developing countries. The question is often what to do to put those transport solutions high on the agenda of municipalities and other policy makers. For example, an analysis of three cities in Africa found that, although experts knew the technical answers to urban transport problems, the cities lacked clear "champions" for transport, and their organizational structure was such that they were unable to respond in a coherent and focused manner (WB SSATP working paper no. 70). The study concluded that what was missing was leadership and coordination, and recommended a reorganization of the way transport is planned and developed in these cities.

However the drive and support for that reorganization still need to come from somewhere, and part of the equation relates to how sustainable transport solutions become a priority in a context of competing problems and agendas of a city.

The globalization of economic activity is one such driver as it has enhanced the role of the city. In developing countries cities are production sites, not only for manufacturing, but also for global services like call centres or back offices. The quality of life offered by these cities can be an important part of their competitive advantage in the globalized economy, with concerns about the risk of traffic injuries or levels of air pollution becoming important in order to attract the right workforce. Although these macroeconomic reasons matter for certain cities, for the vast majority local issues are more likely to influence decision-making.

WHAT ARE THE QUESTIONS THAT CAN TOUCH THE PASSION AND DRIVE THE ACTION OF POLICYMAKERS?

Who will benefit? What types of gains will result from adopting specific transport policies? We argue that it is crucial to make visible and clear the impacts of sustainable transport solutions on a range of tangible benefits, and how they directly affect individuals, their families, their friends and the politicians who represent them. That clarity is not simply a matter of marketing or communication, but a question of using relevant data to evaluate the performance of transport measures, data that can convey issues of central concern to the city population and interests.

The evaluation of the performance of transport policies should involve documentation of all the health impacts those policies are expected to have on the cities resident and visitor populations. There is abundant scientific evidence on the links between transportation and a wide range of risks to health, that will allow the rapid development and testing of the health footprint of transport measures. The communication of those results can speak directly to concerned groups, including globalized businesses in search of lower risks and quality of life for their workforce. The same results can identify health inequalities from transport and the population groups excluded from health protection. The assessment of a "health footprint" from transport should provide a focus for the evaluation of gains from sustainable transport that speaks to the hearts and minds of the people who make up the political makeup of a city. These "footprint" assessments should facilitate the benchmarking of cities' transport performances and become an incentive for cities to adopt and strengthen sustainable transport policies. The footprint assessment should also open the way for communicating more effectively about the gains from sustainable transport policies and measures to city stakeholders and citizens, making those measures more widely under-stood and desired. What is lacking is a concerted effort to measure the health footprint over a range of real-life transport interventions, and to make the results of that initiative widely known. It is high time for this initiative to be developed.

REFERENCES

Andersen LB, Schnohr P, Schroll M, Hein HO (2000). All-cause mortality associ-ated with physical activity during leisure time, work, sports, and cycling to work. *Archives of Internal Medicine*, 160(11):1621–1628.

Aniello C, Morgan K, Busbey A, Newland L (1995). Mapping micro-urban heat islands using Landstat TM and GIS. *Computers and Geosciences*, 21(8):965–969.

Avol EL et al. (2001). Respiratory effects of relocating to areas of differing air pol-lution levels. *American Journal of Respiratory and Critical Care Medicine*, 164(11):2067–2072.

British Medical Association (written by Hillman, M.) (1992). *Cycling: Towards Health and Safety*. Oxford: Oxford University Press.

British Medical Association (1997). *Road Transport and Health*. London: BMA.

Bull FC et al. (2004). Physical inactivity. In: Ezzati M, Lopez AD, Rodgers A, Murray CJL, eds. *Comparative Quantification of Health Risks*. Geneva, Switzerland: World Health Organization.

Buringh E et al. (2002). *On Health Risks of Ambient PM in the Netherlands. Executive Summary*. Bilthoven, Netherlands: National Institute for Public Health and the Environment (RIVM).

Centre for Disease Control and Prevention (1996). *Physical Activity and Health: A Report of the Surgeon General*. Atlanta, GA: US Department of Health and Human Services, Centre for Disease Control and Prevention.

Clancy L, Goodman P, Sinclair H, Dockery DW (2002). Effects of air pollu-tion control on deaths in Dublin, Ireland.: An intervention study. *Lancet*, 360(9341):1210–1214.

Cohen AJ et al. (2004). Urban air pollution. In: Ezzati M, Lopez AD, Rodgers A, Murray CJL, eds. *Comparative Quantification of Health Risks*. Geneva, Switzerland: World Health Organization.

Concha-Barrientos M et al. (2004). Selected occupational risk factors. In: Ezzati M, Lopez AD, Rodgers A, Murray CJL, eds. *Comparative Quantification of Health Risks*. Geneva, Switzerland: World Health Organization.

Dora C, Phillips M, eds. (2000). *Transport, Environment and Health*. WHO Regional Publications, European Series, No 89. Copenhagen, Denmark: World Health Organization, Regional Office for Europe. Available at: www. euro.who.int/document/e72015.pdf.

Dora C, Racioppi F (2003). Including health in transport policy agendas: The role of health impact assessment analysis and procedures in the European experi-ence. *Bulletin of the World Health Organization*, 81(6):399–403.

Ewing R et al. (2003). Relationship between urban sprawl and physical activity, obesity and morbidity. *American Journal of Health Promotion*, 18(1):47–57.

Frumkin H (2002). Urban sprawl and public health. *Public Health Reports*, 117(3):201–217.

Gubler DJ, Meltzer M (1999). Impact of dengue/dengue haemorrhagic fever on the developing world. *Advances in Virus Research*, 53:35–70.

Hajat S, Armstrong BG, Gouveia N, Wilkinson P (2005). Mortality displacement of heat related deaths: A comparison between Delhi, Sao Paulo and London. *Epidemiology*, 16(5):613–620.

Hedley AJ et al. (2002). Cardiorespiratory and all-cause mortality after restrictions on sulphur content of fuels in Hong Kong: An intervention study. *Lancet*, 360(9346):1646–1652.

Hillman M (1993). Cycling and the promotion of health. *Policy Studies*, 14(2):49–58.

Hoek G et al. (2002). Association between mortality and indicators of traffic-related air pollution in the Netherlands: A cohort study. *Lancet*, 360(9341):1203–1209.

Hou L et al. (2004). Commuting physical activity and risk of colon cancer in Shanghai, China. *American Journal of Epidemiology*, 160(9):860–867.

Ice Interface for Cycling Expertise & the Association of Dutch Municipalities VNG (2000). *The Economic Significance of Cycling: A Study to Illustrate the Costs and Benefits of Cycling Policy*. The Hague, Netherlands: Habitat Platform Foundation. Available at: www.i-ce.info/download/publications/Ec_Significance_eng.pdf, accessed 30 October 2005.

Idso CD, Idso SB, Balling RC (2001). An intensive two week study of an urban CO_2 dome in Phoenix, Arizona, USA. *Atmospheric Environment*, 35(6):995–1000.

International Panel on Climate Change, IPCC (2001a). Climate Change 2001. Synthesis Report. Cambridge, UK: WMO/UNEP.

International Panel on Climate Change, IPCC (2001b). Climate Change 2001. Mitigation: Contribution of Working Group III to the Third Assessment Report. Cambridge, UK: Cambridge University Press.

International Panel on Climate Change, IPCC (2001c). Climate Change 2001. The Scientific Basis: Contribution of Working Group I to the Third Assessment Report. Cambridge, UK: Cambridge University Press.

Kenworthy J, Laube F (2002). Urban Transport patterns in a global sample of cities & their linkages to transport infrastructures, land use, economics & environment. *World Transport Policy and Practice*, 8(3):5–20.

Krzyzanowski M, Kuna-Dibbert B, Schneider J., eds. (2005). *Health Effects of Transport-Related Air Pollution*. Copenhagen, Denmark: World Health Organization, Regional Office for Europe.

McMichael AJ, Githeko A (2001). Human health. In: McCathy JJ, et al., eds. *Climate Change 2001: Impacts, Adaptation and Vulnerability*. Cambridge, UK: Cambridge University Press, pp 451–485.

McMichael et al. (2004). Climate change. In: Ezzati M, Lopez A, Rodgers A, Murray C, eds. *Comparative Quantification of Health Risks: Global and Regional Burden of Disease Due to Selected Major Risk Factors*. Geneva, Switzerland: World Health Organization.

Mohan D, Tiwari G (1999). Linkages between environmental issues, public transport, non-motorised transport and safety. *Economic and Political Weekly*, XXXIV(25):1580–1596.

Mohan D, Tiwari G (2000). Mobility, environment and safety in megacities: Dealing with a complex future. *IATSS Research*, 24(19):39–46.

Nantulya VM, Reich MR (2002). The neglected epidemic: Road traffic injuries in developing countries. *BMJ*, 324(7346):1139–1141.

O'Neill B, Mohan D (2002). Reducing motor vehicle crash deaths and injuries in newly motorising countries. *BMJ*, 324(7346):1142–1145.

O'Neill MS et al. (2003). Health, wealth and air pollution: Advancing theory and methods. *Environmental Health Perspectives*, 111(16):1861–1870.

Odense Municipality (Odense Kommune) (2004). *Evaluering af Odense-Danmark's Nationale Cykelby [Evaluation of Odense, Denmark's National Cycle City]* (in Danish with English summary). Odense, Denmark: Odense Municipality.

OECD Environment Outlook (2001). Paris: Organization for Economic Cooperation and Development.

Peden M et al., eds. (2004). *The World Report on Road Traffic Injury Prevention*. Geneva, Switzerland: World Health Organization.

Peden M, McGee K, Sharma G (2002). *The Injury Chart Book: A Graphical Overview of the Global Burden of Injuries*. Geneva, Switzerland: World Health Organization.

Pope CA et al. (2002). Lung cancer, cardiopulmonary mortality, and long-term exposure to fine particulate matter air pollution. *JAMA*, 287:1132–1141.

Prüss-Üstün A, Corvalàn C (2005). *Estimating How Much Disease Is Caused by the Environment*. Geneva, Switzerland: World Health Organization.

Prüss-Üstün A, Fewtrell LJ, Landrigan P, Ayuso-Mateos JL (2004). Lead exposure. In: Ezzati M, Lopez AD, Rodgers A, Murray CJL, eds. *Comparative Quantification of Health Risks*, Volume 1. Geneva, Switzerland: World Health Organization.

Qin HL et al. (2004). Effect of environment on extremely severe road traffic crashes: Retrospective epidemic analysis during 2000–2001. *Chinese Journal of Traumatology*, 7(6):323–329.

Racioppi F, Dora C, Rutter H (2005). Urban settings and opportunities for healthy lifestyles: Rediscovering walking and cycling and understanding their health benefits. *Built Environment*, 31(4):302–314.

Roemer WH, van Wijnen JH (2001). Daily mortality and air pollution along busy streets in Amsterdam, 1987–1998. *Epidemiology*, 12(6):649–653.

Rutter H (2005). Personal communication.

Sælensminde K (2002). Walking and cycling track networks in Norwegian cities. Cost benefit analyses including health effects and external costs of road traffic. TOI Report 567 (in Norwegian with English summary). Oslo, Norway: Institute of Transport Economics.

Sanderson E et al. (2005). Human exposure to transport-related air pollution. In: Krzyzanowski M, Kuna-Dibbert B, Schneider J, eds. *Health Effects of Transport-Related Air Pollution*. Copenhagen, Denmark: World Health Organization, Regional Office for Europe.

Schwartz J, Laden F, Zanobetti A (2002). The concentration-response relationship between PM 2.5 and daily deaths. *Environmental Health Perspectives*, 110(10):1025–1029.

Thomas VM et al. (1999). Effects of reducing lead in gasoline: An analysis of the international experience. *Environmental Science and Technology*, 33(22):3942–3948.

Tiwari G (2003). Transport and land-use policies in Delhi. *Bulletin of the World Health Organization*, 81(6):444–450.

WHO (2002). *World Health Report 2002. Reducing Risks, Promoting Healthy Life*. Geneva, Switzerland: World Health Organization. Available at: www.who.int/whr/2002/, accessed March 2005.

WHO (2005). *Preventing Chronic Diseases: A Vital Investment*. Geneva, Switzerland: World Health Organization.

WHO Regional Office for Europe (2006). *Health Effects and Risks of Transport Systems: The HEARTS Project*.

Zhou L et al. (2004). Evidence for a significant urbanization effect on climate in China. *Proceedings of the National Academy of Sciences of the United States of America*, 101(26):9540–9544.

Chapter 17

Traffic safety, city structure, technology and health

Dinesh Mohan

CONTENTS

INTRODUCTION

Road safety

Road traffic injuries (RTIs) are the only public health problem where society and decision makers still accept death and disability on a large scale among young people. This human sacrifice is deemed necessary to maintain high levels of mobility and is seen as a necessary "externality" of doing business. Discussion only revolves around the number of deaths and injuries we are willing to accept. The sole departure from this mode of thinking is the "Vision Zero" that originated in Sweden. In October 1997, the Road Traffic Safety Bill founded on Vision Zero was passed by a large majority in the Swedish parliament. The Vision Zero is that eventually no one will be killed or seriously injured within the road transport system. However, even in Sweden, believers are few.

Deaths among workers in factories, mines, railroads and dockyards were commonplace and accepted in the early twentieth century. This is not acceptable anymore. Many societies do not sanction the death penalty, no matter how serious the crime. About a hundred deaths caused by the recent spread of the severe acute respiratory syndrome (SARS) mobilised international efforts to arrest the disease. Millions of demonstrators came out on the streets to protest a war in their belief that nothing justifies the deaths of innocent individuals. But this attitude is absent when it comes to road traffic injuries.

Recent estimates suggest that RTIs result in more than one million fatalities worldwide annually. A vast majority of these deaths include people who are less than 50 years old. Another 20–30 million people suffer injuries that need hospitalisation or expert medical treatment. It is not that individuals and decision makers are not concerned about the epidemic of RTIs. Governments in high-income countries (HICs) have spent a great deal of effort in establishing road safety agencies and standards as well as some on research. On the other hand, governments in low- and middle-income countries (LMIC) have shown concern for more than two decades but have not established effective agencies or spent any money on research. However, the attitudes of both HIC and LMIC governments are similar – RTIs and deaths are here to stay, we can only attempt to reduce them. Citizens of LMICs, however, have shown a greater intolerance of deaths caused by road traffic crashes. In many Asian and African countries, hardly a day goes by when an angry crowd does not try to lynch a driver or burn a vehicle involved in a pedestrian crash. Villagers on their own have also constructed "illegal" speed bumps (speed breakers) in thousands of villages to slow down vehicles speeding through their neighbourhoods.

So where does the problem lie? The main problem is the common and persistent fixation on fault. If human error is seen as the root cause of RTIs, it follows that the problem must be dealt with by educating road users. It also follows that those who do "wrong" may suffer injuries. And this mindset has continued in the face of all scientific evidence that educating road users is not the most productive way to reduce RTIs.

A second problem is that work on road safety is still not recognised as a scientific occupation in our academic institutions and among decision makers. This results in a huge turnover in "experts" who work in this area. Very few work on road safety as their dominant area of activity. So each batch of experts goes on rediscovering or reinventing the same wheel, repeatedly. It has been known for over 25 years that driver education does not significantly reduce the incidence of RTIs. But we must patiently go through the same process over and over again, as each new batch of "experts" ultimately sees the light and stops focusing on education as the main solution to RTIs.

An offshoot of this education fixation is the dogged belief among road safety consultants that people in poorer countries especially need road traffic education. This equation of underdeveloped with "needing education" flies in the face of research findings and simple logic. Not a single scientific study from an LMIC has demonstrated a correlation between lack of "knowledge" and RTI rates. Age-specific RTI fatality rates for children (0–14 years) are very similar in HICs. Yet, the first advice of international consultants working in LMICs is to start a children's road safety education programme.

The consequence of all this is that essential policies and countermeasures needed to control the epidemic of road traffic injuries remain neglected. Matthijs Koornstra highlights this indirectly by showing through his mathematical models how long it will take for most countries to reverse the trend of rising death and injury rates. These black-box models are based on past experience and, as of now, are pretty good at predicting the future. They depend on interactions between average time constants for technology development and societal responses for facing crises. The important question is whether there is any possibility of these time constants reducing in the coming years. If the answer is "no", then we are in trouble.

Transportation and health

Carlos Dora makes a strong case that the motor car may be one of the most health-destroying technologies that we have allowed to proliferate in the past one hundred years. And quite willingly at that. If we consider the dire predictions of the effects of global warming, then we do indeed have a problem at hand. Hermann Knoflacher seems to believe that speed is a drug more addictive than cocaine and the development of the car in a speck of evolutionary time has overwhelmed our evolutionary capabilities. If Knoflacher is right then it is difficult to believe that the negative health effects of car use can be reversed easily. A summary of these deleterious health effects (short-term) is shown in Figure 17.1. Out of all these concerns, traffic management experts and road builders focus mainly on congestion mitigation efforts, mostly unsuccessfully. However, congestion reduction has wide political and business support as it involves the building of more roads and infrastructure. This in turn worsens the greenhouse gas emissions, noise production and separation of communities in cities.

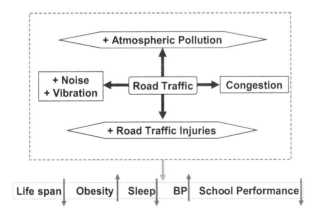

Figure 17.1 Health effects of road transport.

There is widespread political, professional and community agreement for the reduction of pollution, but we end up focusing only on reduction of exhaust (tailpipe) emissions. Socially and technically this is easier to push as it does not involve changes in lifestyles and city infrastructure. But if Carlos Dora's list of ill health effects is to be taken seriously, then tailpipe issues are not enough. The following adverse health effects are not resolved with a reduction in emissions from vehicles:

- Noise from wide and busy roads – reduction in sleep and concentration, especially among children and the elderly. This also results in higher blood pressure among the elderly and learning difficulties among children.
- Obesity – due to lack of exercise, causes a reduction in life span.
- Separation of communities by wide busy roads – this results in less social interaction, a reduction in walking and bicycling and a lack of community development.

All of the above threaten sustainable futures both directly and indirectly. If these issues are so important, then it is important to understand the role of city structure and transportation choices made by people. The issue that we need to clarify is whether people are being forced to live in ways that are detrimental to their health, or they are content with the current forms of living and are willing to pay these health costs for the same.

ROAD SAFETY – NEED FOR A MORE ROBUST SCIENCE

Koornstra makes a shopping list of conditions under which road safety progress can be made. Almost no safety professional would disagree with these proposals. However, he does not discuss the conditions under which these solutions find wider applicability and those when they do not. He also takes for granted the provision of road safety in itself as a public good, which is correct, but does not elaborate the need for road safety as a necessary precondition for reducing greenhouse gases and other pollutants from motor vehicles.

Road safety and cleaner air

Most of the megacities in the world are already located in LMICs and many more cities in these countries will grow to populations of 10 million or more in the next few decades (World Health Organisation, 1998). All these cities are faced with serious problems of inadequate mobility and access, vehicular pollution, road traffic crashes and crime on their streets. The increasing use of cars and motorised two-wheelers add to these problems and this trend does not seem to be abating anywhere. Many recent reports suggest that improvements in public transport and promotion of non-motorised modes of transport can help substantially in alleviating some of these problems (Mohan et al., 1996; Yong and Xiaojiang, 1999; OECD, 2000; European Commission, 2001). However, LMIC cities have very mixed land-use patterns, and a very large proportion of all trips are walking or bicycle trips; of the motorised trips, more than 50% are by public transport or shared paratransit modes; compared to HICs, trips per capita per day are lower and a significant proportion of trips can be less than 5 km in length; and the costs of motorised travel are high compared to average incomes (Mohan et al., 1996). In spite of these structural advantages, the air pollution levels in LMIC cities remain high. What these cities do not have are efficient public bus systems, safe and convenient walkways and bicycle lanes, the best in fuel quality and vehicle technology, and strict and efficient vehicle maintenance systems. However, improvements in these will take time, large financial investments and may be difficult to implement for a variety of reasons.

In addition to the problems of pollution, deaths and injuries due to road traffic crashes are also a serious problem in LMICs. According to one estimate, the losses due to accidents in LMICs may be comparable to those due to pollution (Vasconcellos, 1999). These problems become difficult to deal with because there are situations in which there are conflicts between safety strategies and those that aim to reduce pollution (OECD, 1997). For example: smaller and lighter vehicles can be more hazardous but they are less energy-consuming; congestion reduces the probability of serious injury due to crashes but increases pollution; and an increase in bicycling rates can decrease pollution, but may increase crashes if appropriate facilities are not provided. Such issues make transport planning in LMIC cities a very complex affair. HIC cities have not experienced the existence of such a large proportion of motorised two-wheelers, paratransit vehicles and non-motorised modes of transport sharing road space with cars and buses.

Walking and bicycling are the only clean modes of transport available. The use of these modes reduces as incomes rise and cities become unfriendly to these modes, designing roads with only motor vehicles as a priority. Pan Haixiao tells us how these modes are being discouraged even in China with policies that favour car use. Along with this headlong rush for car-oriented planning, all policymakers assure us that they want the use of public transport to increase and that they are investing funds on public transport projects. But these policies to promote the use of public transport are unlikely to reduce the emission of greenhouse gases because many walking and bicycle trips may be converted to fossil fuel-using public transport trips without an adequate shift from cars to public transport.

The main reason for this is that walking is not safe. Even in a safety-conscious city like Copenhagen bus commuters are not as safe as pedestrians. Figure 17.2 shows us that bus commuters are much more likely to get killed during the pedestrian parts of their access trips than during the

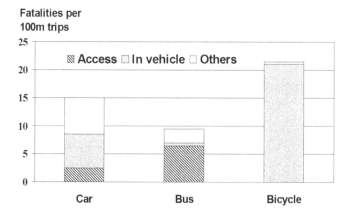

Figure 17.2 Trip types and fatality rates in central Copenhagen.

bus journeys themselves (Jorgensen, 1996). This is probably true the world over. The high risks of injuries and fatalities in urban areas to pedestrians, bicyclists and commuters in access trips have been documented from all over the world. The greatest risk to schoolchildren from bus-related injuries was found to be as pedestrians after alighting from a bus in New South Wales, Australia (Cass et al., 1997). In Mexico City, 57% of deaths from traffic crashes involve pedestrians (Híjar et al., 2001). Injury to pedestrians was the most frequent cause of multiple trauma (54%) among children 0–16 years in a large Spanish urban area (Sala et al., 2000). In California a motor vehicle versus pedestrian accident study reported that these accidents are common and the high mortality rate among the elderly indicated the need for more aggressive and effective prevention efforts (Peng and Bongard, 1999). A study from Canada showed that children's exposure to traffic (number of streets crossed) and injury rates were positively correlated (Macpherson et al., 1998). In Kumasi, Ghana, the most common causes of injury (40.0%) to children were pedestrian knock-downs (Abantanga and Mock, 1998). A study of older people's lives in the inner city in Sydney, Australia, showed that environmental hazards, such as pedestrian safety and traffic management, affect the whole population and require intervention at government level (Russell et al., 1998). A study from Seattle showed that 66% of the fatal injuries occurred on city or residential streets, 29% occurred on major thoroughfares, and a single urban highway accounted for 12% of pedestrian fatalities and represented a particularly hazardous traffic environment (Harruff et al., 1998).

Quite obviously, people's fears regarding safety on the roads when using public transport are not unjustified. A large proportion of the decrease in road traffic injuries and deaths in HICs is the result of the availability of cars that provide much greater safety to the occupants in crashes, and the result of a very significant reduction of the presence of pedestrians and cyclists on HIC streets. Recent estimates from the United Kingdom suggest that the number of trips per person on foot fell by 20% between 1985/86 and 1997/99 (Select Committee on Environment and Regional Affairs, 2001). Since every public transport trip involves two non-motorised access trips, the use of public transport is not likely to increase significantly unless commuter safety is ensured.

Public transport and overall safety

Most policymakers assume that increasing use of public transport can result in a reduction in RTIs because they only take into account the risks of bus or metro occupants and not of the access trips. However, the preliminary results of an analysis by the Transportation Research and Injury Prevention Programme (TRIPP) for the present situation in Delhi indicates that increased bus use in the two mode-shift scenarios considered (car and motorcycle users shifting to buses) leads to higher traffic fatalities because it involves substituting low-risk travel patterns with ones that have higher risks.[1] In each scenario, the increased bus use results in a larger number of pedestrians (due to bus-stop access trips) and buses (to cater to the higher bus-use demand), the threat–victim pair that has one of the highest fatality risks per unit exposure. In the first case, merely shifting 5% of car occupants to buses causes an additional 66 pedestrian deaths due to crashes with buses and trucks. In the second case, the shift from motorcycles to bus-use reduces motorcycle deaths by 50 but increases pedestrian deaths by 68.

Public transit systems have a strong appeal for transportation engineers because they contribute less pollution and congestion, and provide an egalitarian solution to the mobility needs of a city. However, the perception that they result in increased safety may not always be correct. Our analysis of current road traffic injury statistics in Delhi found that buses were involved in fatal crashes in numbers that were disproportionate to their use. As a result, increasing bus use would result in increasing traffic fatalities, primarily among pedestrians, the most vulnerable road users.

There is little doubt that a sustainable transportation system for cities in LMICs will necessarily rely on buses. However, under existing conditions, promoting bus travel will only result in a higher road traffic injury death toll unless appropriate technology and infrastructure for bus safety are provided. Furthermore, reliance on bus use implies high levels of walking and bicycling. Such a transport system can be feasible only if the streets are safe from crime and provide public amenities, such as shops, restaurants and street vendors, and safe access trips. None of these conditions will exist without appropriate investment in suitable infrastructure. Preliminary results from our study suggest that the successful promotion

---

[1] Buses and Urban Road Safety. Kavi Bhalla, Geetam Tiwari, Dinesh Mohan, Majid Ezzati, and Ajay Mahal (personal communication)

of public transport will depend on the existence of safe walking and bicycling infrastructure as a precondition. Ultimately, even the possibility of having cleaner air in our cities also depends on a safer city, and this may depend on the structure of cities as much as on safer road design and traffic management.

Cities and their road safety experience

Figure 17.3 shows road traffic fatalities per million population of a number of cities around the world. These data show that there are wide variations across income levels and within similar income levels. The risk varies by a factor of about 20 between the best and the worst cities. Some characteristics are summarised below:

- The highest fatality rates seem to be experienced by cities in the mid-income range of USD 2,000–10,000 per person per year.
- Overall fatality risk in cities with very low per capita incomes (less than USD 1,000) and those with high incomes (greater than USD 10,000) seem to be similar.
- There is a great deal of variation even in those cities where the per capita income is greater than USD 20,000 per year.

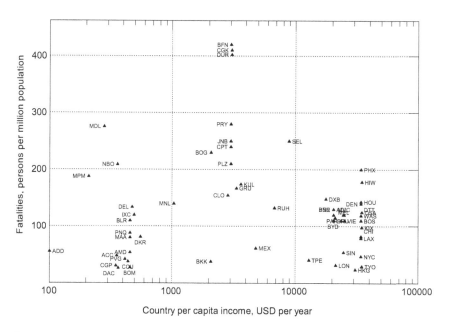

Figure 17.3 Fatality risk in traffic crashes by city (cities identified by airport code).

These patterns appear to indicate that it is not enough to have the safest vehicle technology in high-income countries to ensure low road traffic fatality rates uniformly across cities in those locations. Even in very low-income countries, the absence of funds and possibly unsafe roads and vehicles does not mean that all cities have high overall fatality rates.

Figures 17.4 and 17.5 show pedestrian and motorcycle fatality rates per million population in cities in different countries respectively.

The provision of safely designed roads and modern safe vehicles may be a necessary condition for low road fatality rates in cities but it is not a sufficient one. The fact that there are wide variations for overall fatality rates among high-income cities, where the availability of funds, expertise and technologies are similar, indicates that other factors, such as land-use patterns and exposure (distance travelled per day, presence of pedestrians, etc.) also play a very important role. This is probably why many European cities tend to have lower rates than those in the United States.

Vehicle speed is very strongly related to both the probability of a crash and the severity of injury – a 1% increase in average speeds can result in 3–4% increase in fatalities (Peden et al., 2004). This may be the reason why some cities in middle-income countries have high fatality rates, because they have higher vehicle ownership than low-income countries and roads encouraging unsafe speeds without adequate attention being given to road safety. Similarly cities that are considered to have greater traffic congestion (hence lower speeds) have higher rates than those with less congestion

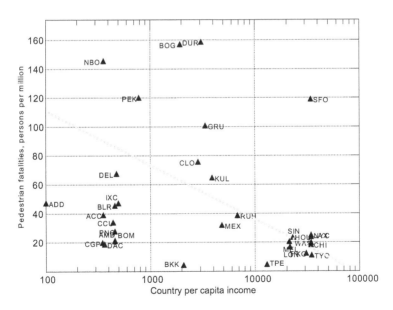

Figure 17.4 Pedestrian fatality risk in traffic crashes by city (cities identified by airport code).

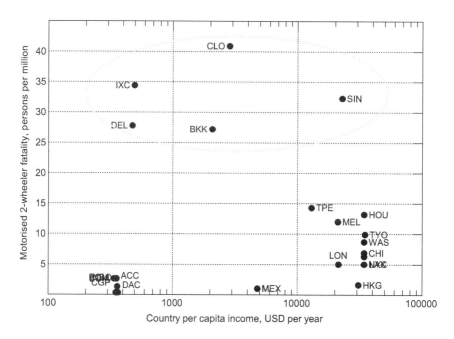

Figure 17.5 Motorcyclist fatality risk in traffic crashes by city (cities identified by airport code).

even though their incomes may be similar – Mumbai in India with higher congestion has lower rates than Delhi in India with lower congestion levels, and New York in the United States has a lower rate than Houston, also in the United States. However, motorcycle rates do not seem to depend on income levels at all. This may be due to the high risk faced by motorcycle riders and so wherever motorcycles have a high modal share there will be high fatality rates.

Fatality rates and city structure

The discussion above indicates that RTI fatality rates do not depend upon road and vehicle design alone and may be also depend on exposure, income, traffic management and policing, which in turn also depend on society income levels. To eliminate the issue of income we decided to examine the experience of cities within the United States. We have selected the United States because researchers there generate the largest amount of scientific information on road safety and society is not much worse off than any other in terms of the availability of funds for road safety. Figure 17.6 shows the pedestrian and motor vehicle fatality rates per 100,000 population in 245 cities (population > 100,000 persons) in the United States (Shankar, 2003). Table 17.1 shows 10 cities that have zero pedestrian deaths, of which eight also have no motor vehicle deaths. All these cities have populations

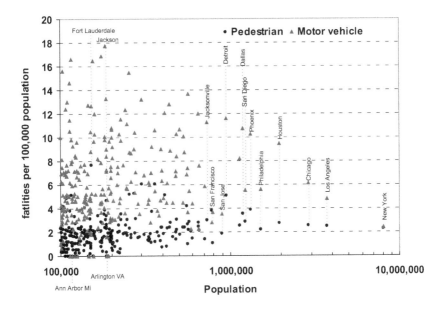

Figure 17.6 Pedestrian and motor vehicle fatality rates per 100,000 population in 245 cities (population > 100,000 persons) in the United States of America.

Table 17.1   Ten cities with lowest pedestrian fatality rates in the United States

City	Population	Pedestrian fatality per 100,000 population	Motor vehicle fatality per 100,000 population
Ann Arbor, MI	114,024	0	0
Spring Valley CDP, NV	117,390	0	0
Elizabeth, NJ	120,568	0	0
East Los Angeles CDP, CA	124,283	0	0
Metairie CDP, LA	146,136	0	0
Sunrise Manor CDP, NV	156,120	0	0
Paradise CDP, NV	186,070	0	0
Arlington CDP, VA	189,453	0	0
Bellevue, WA	109,569	0	3.04
Manchester, NH	107,006	0	4.98

between 100,000 and 200,000. This is in contrast with 13 cities in the same population range that have some of the highest rates in the country (Table 17.2).

Table 17.3 shows data for eight cities in the United States that have the highest pedestrian fatality rates, and most of them also have very high motor vehicle fatality rates. The populations of these cities vary from among the lowest in the group (Columbia, SC) to almost 1 million (Detroit, MI).

Table 17.2   Thirteen cities in the United States (population < 200,000) with pedestrian fatality rates > 3 per 100,000 population

City	Population	Pedestrian fatality per 100,000 population	Motor vehicle fatality per 100,000 population
Gary, IN	102,746	4.22	15.57
Waterbury, CT	107,271	3.73	12.43
Beaumont, TX	113,866	3.81	14.63
Columbia, SC	116,278	5.16	16.63
Fayetteville, NC	121,015	3.31	7.16
Pomona, CA	149,473	3.12	4.91
Fort Lauderdale, FL	152,397	7.66	12.68
Dayton, OH	166,179	3.41	8.22
Reno, NV	180,480	3.51	4.06
Salt Lake, UT	181,743	3.85	10.09
Jackson, MS	184,256	3.08	17.72
San Bernardino, CA	185,401	3.42	8.99
Orlando, FL	185,951	4.66	13.27

Table 17.3   Cities in the United States with pedestrian fatality rates > 5 per 100,000 population

City	Population	Pedestrian fatality per 100,000 population	Motor vehicle fatality per 100,000 population
Detroit, MI	951,270	5.05	11.59
Atlanta, GA	416,474	5.12	12.09
Columbia, SC	116,278	5.16	16.63
Louisville, KY	256,231	5.2	15.48
Newark, NJ	273,546	5.24	9.02
Tampa, FL	303,447	6.04	13.18
Miami, FL	362,470	6.07	10.58
Fort Lauderdale, FL	152,397	7.66	12.68

It is clear that in US cities, size is not a determining factor for pedestrian or motor vehicle fatality rates. However, it is interesting that New York, which is the largest city in the United States, has much lower vehicle and pedestrian crash rates than most cities in the country. If we take the five largest cities in the country – Philadelphia, PA, Houston, TX, Chicago, IL, Los Angeles, CA, and New York, NY (Figure 17.6) – we find that all of them have relatively low pedestrian fatality rates, but, other than New York, they have high motor vehicle fatality rates. For motor vehicle fatality rates to be high, there have to enough crashes where the effective impact velocity is higher than 80–100 km/h for belted and airbag-protected occupants. The fact that vehicle fatality rates are low in New York, means that average speeds would also be low there. The same explanation should hold for the eight cities that have vehicle fatality rates of zero per 100,000 population and also have rates of zero for pedestrians (Table 17.1). The fact that a city can have high fatality rates for vehicles (indicating higher average vehicle speeds) and low rates for pedestrians implies that these cities would have low pedestrian exposure and hence lower rates.

In the absence of detailed traffic modal share, speed and crash information, we can only make informed guesses on what is happening in all these cities. The international data show that per capita income is not the only determining criterion for fatality rates, as the rates can vary by a factor of three among the richest nations. To control for different vehicle design and road design policies, we have compared rates for cities within the United States. These data show that within the same state:

- Cities have different crash rates – San Diego and San Jose in California.
- Cities have different patterns – San Francisco has a higher rate for pedestrians and Los Angeles has a higher rate for vehicles.
- One city can have a zero rate of fatalities (East Los Angeles CDP, CA) and another city in the same state with a similar population can have one of the highest rates (San Bernardino, CA).

We also know that improvements in the crashworthiness of vehicles and the use of seat belts, airbags and other safety devices can reduce fatality rates by 30–70%, alcohol control by about 30%–40% (Peden et al., 2004), and that enhancement in road and infrastructure facilities can lead to an increase in fatalities (Noland, 2003). But, the differences in rates across cities in the United States and internationally show differences that vary by factors of three or more. This seems to suggest that city structure, modal share split and the exposure of motorists and pedestrians may have a greater role in determining fatality rates than vehicle and road design alone.

Therefore the results from the analysis of city data seem to suggest very strongly that cities with high motor vehicle fatality rates must be those where exposure and speeds of motorists are high, and pedestrian fatality rates can be low if pedestrian exposure is low. Within the United States, where incomes, availability of technology, knowledge, road design and vehicle specifications can all be similar across cities, these differences in speeds and exposure are probably accounted for by the structure of the cities.

Figure 17.7 shows the street layout in Ann Arbor, MI, which has a zero fatality rate and Columbia, SC, which has a very high fatality rate. Both the maps are printed on the same scale. It appears that Ann Arbor has a denser street network and fewer wide avenues. If this is true, then this might explain why speeds may be lower on narrower streets with frequent intersections in Ann Arbor compared with Columbia. Ann Arbor's dense street network may also encourage walking and bicycling trips.

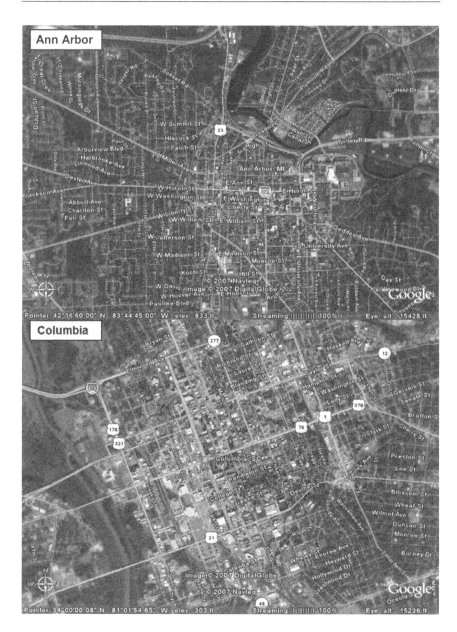

Figure 17.7 Aerial images 5 km in width of Ann Arbor, MI (*top*) and Columbia, SC (*bottom*).

At the international level, cities that have modernised and expanded in the past few decades are those in the per capita income range of USD 1,000–10,000, and these are the cities with very high fatality rates (Durban, Johannesburg, Tehran, etc.). Typically, these cities have built wide avenues and high-speed corridors within the city. In India, Delhi has a high fatality rate, and Mumbai and Kolkata low rates. Here also, Delhi has much faster vehicle traffic than Mumbai and Kolkata and a lower density of through traffic streets. Figure 17.8 shows the street layout in London and Delhi on the same scale. Delhi seems to have a similar area devoted to road space on the whole as London. But residential developments in Delhi seem to be larger and consequently the distances between through traffic streets larger than London. This puts pressure on authorities in Delhi to have much wider arterial roads. This encourages high speeds during off-peak hours, resulting in high pedestrian and bicycle crash rates. High pedestrian and bicycle fatality rates discourage the use of non-motorised modes and the use of public transport. In addition large residential developments lead to longer access trips for public transport, also discouraging its use. Clearly, city street structure has a very strong influence whether sustainable transport policies can succeed or not.

CONCLUSIONS

- Adverse health effects of road transport are caused by the following factors:
- Noise leading to lack of sleep and increase in blood pressure among the elderly and adverse learning outcomes among children
- Emissions such as SO_2, NO_x, benzene, ozone and other organic compounds
- Carbon dioxide and associated global warming
- Morbidity and mortality due to road traffic crashes
- Obesity due to lack of exercise
- Reduction in community interaction leading to psychological pressures and adverse social outcomes especially among the young and the elderly.

Issues for sustainable transport:

- In the next ten years purely technological solutions in the design of vehicles and their engines can lead to a small reduction in injury and fatality rates and a significant reduction in emissions such as SO_2, NO_x, benzene, ozone and other organic compounds.
- Carbon dioxide emissions are not likely to be reduced by current technological options.

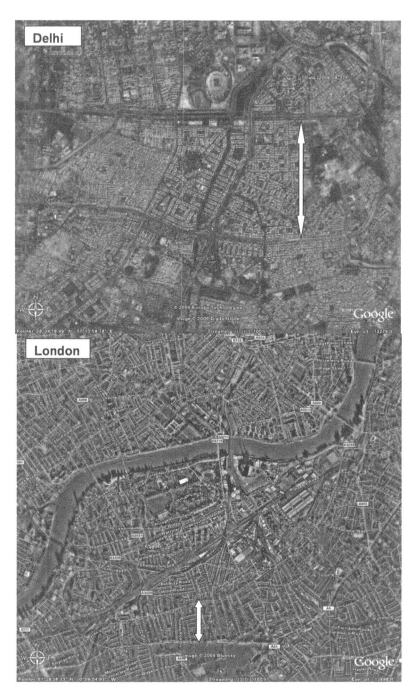

Figure 17.8 Aerial images of Delhi (*top*) and London (*bottom*) on the same scale. Arrows show distances between through traffic streets.

- Currently, no significant noise reduction can be predicted by technical design solutions.
- Policing and traffic management techniques are not likely to reduce traffic fatality rates by more than 50–75%, though we see a difference of fatality rates by factors of 4–10 between cities.
- Obesity issues will not be affected by vehicle or road design.
- Community interaction issues will not be affected by vehicle design.

Therefore, it seems that if we are to promote walking, bicycling and public transport use, we will have to make traffic safety a priority along with city structure designs that incorporate the following:

- Street design ensuring the safety of non-motorised modes.
- Vehicle speed control by street design and ultimately intelligent traffic system (ITS) control on vehicles.
- A denser layout of through traffic streets with narrower cross sections.
- Smaller size of residential neighbourhoods.

REFERENCES

Abantanga, F.A., and Mock, C.N. (1998) Childhood Injuries in an Urban Area of Ghana. A Hospital-Based Study of 677 Cases. *Pediatric Surgery International* **13**, 515–518.

Cass, D.T., Ross, F., and Lam, L. (1997) School Bus Related Deaths and Injuries in New South Wales. *Medical Journal of Australia* **166**, 107–108.

European Commission. (2001) *European Transport Policy for 2010: Time to Decide*. White Paper. Luxembourg: Office for Official Publications of the European Communities.

Harruff, R.C., Avery, A., and Alter-Pandya, A.S. (1998) Analysis of Circumstances and Injuries in 217 Pedestrian Traffic Fatalities. *Accident Analysis and Prevention* **30**, 11–20.

Híjar, M.C., Kraus, J.F., Tovar, V., and Carrillo, C. (2001) Analysis of Fatal Pedestrian Injuries in Mexico City, 1994–1997. *Injury* **32**, 279–284.

Jorgensen, N.O. (1996) The Risk of Injury and Accident by Different Travel Modes. In: *International Conference on Passenger Safety in European Public Transport* pp. 17–25. Brussels, Belgium: European Transport Safety Council.

Macpherson, A., Roberts, I., and Pless, I.B. (1998) Children's Exposure to Traffic and Pedestrian Injuries. *American Journal of Public Health* **88**, 1840–1843.

Mohan, D., Tiwari, G., Saraf, R., Kale, S., Deshmukh, S.G., Wadhwa, S., and Soumitri, G. Delhi on the Move: Future Traffic Management Scenarios. 1996. In: *Transportation Research and Injury Prevention Programme*, Delhi, India. Indian Institute of Technology.

Noland, R.B. (2003) Traffic Fatalities and Injuries: The Effect of Changes in Infrastructure and Other Trends. *Accident; Analysis and Prevention* **35**, 599–611.

OECD. (1997) *Integrated Strategies for Safety and Environment.* Paris: Organisation for Economic Co-operation and Development.

OECD. (2000) *Synthesis Report on Environmentally Sustainable Transport: Futures, Strategies and Best Practices.* Vienna, Austria: Austrian Ministry of Agriculture, Forestry, Environment and Water Management.

Peden, M., Scurfield, R., Sleet, D., Mohan, D., Hyder, A.A., Jarawan, E., and Mathers, C. (2004) *World Report on Road Traffic Injury Prevention.* Geneva, Switzerland: World Health Organization.

Peng, R.Y., and Bongard, F.S. (1999) Pedestrian Versus Motor Vehicle Accidents: An Analysis of 5,000 Patients. *Journal of the American College of Surgeons* **189**, 343–348.

Russell, C., Hill, B., and Basser, M. (1998) Older People's Lives in the Inner City: Hazardous or Rewarding? *Australian and New Zealand Journal of Public Health* **22**, 98–106.

Sala, D., Fernández, E., Morant, A., Gascó, J., and Barrios, C. (2000) Epidemiologic Aspects of Pediatric Multiple Trauma in a Spanish Urban Population. *Journal of Pediatric Surgery* **35**, 1478–1481.

Select Committee on Environment and Regional Affairs. *Walking in Towns and Cities. Eleventh Report.* (2001) London: House of Commons.

Shankar, U. (2003) *Pedestrian roadway fatalities.* National Center for Statistics and Analysis, National Highway Traffic Safety Administration, Washington, DC, 1–59.

Vasconcellos, E.A. (1999) Urban Development and Traffic Accidents in Brazil. *Accident Analysis and Prevention* **31**, 319–328.

World Health Organisation. (1998) *Healthy Cities: Air Management Information System.* AMIS 2.0 CD. Geneva, Switzerland: World Health Organization.

Yong, W., and Xiaojiang, L. (1999) Targeting Sustainable Development for Urban Transport. *Workshop on Urban Transport and Environment.* Beijing, China: CICED.

Chapter 18

Paratransit, taxis and non-motorised transport

A review of policy debates and challenges

Roger Behrens

CONTENTS

INTRODUCTION

Background

The brief for this chapter (originally a discussion paper) was to synthe-sise current policy debates and challenges relating to the facilitation and management of 'paratransit, taxis and non-motorised transport' in both 'developing world' and 'developed world' city contexts, and to suggest ways forward with respect to appropriate responses in policy and practice. Specific questions raised for attention related to the preservation and increase of non-motorised transport (NMT) mode share, the resolution of prevailing NMT infrastructure design and safety problems, the improvement of for-hire and paratransit services, and appropriate linkages between these modes and scheduled public transport services.

'Paratransit, taxis and non-motorised transport' has been interpreted to encompass an overlap between non-private motorised transport, unscheduled public transport, and non-motorised passenger modes. More detailed definitions will be offered at the beginning of subsequent sections.

Purpose, scope and limitations

The purpose of the chapter is, in essence, to report on the findings of a selective review of key literature, and to extract and synthesise key policy debates.

The potential scope of the chapter is vast, incorporating numerous forms of private non- motorised, for-hire and unscheduled paratransit modes (see Figure 18.1). The breadth of the subject field necessitates that the material covered is neither definitive nor comprehensive. The literature reviewed and material presented has been selected on the basis of relevance to the overarching workshop theme of sustainable transport.

The subject matter is limited to urban land passenger transport modes, and, within this broad categorisation, only selected modes are discussed. While of unquestionable importance, some policy debates surrounding modes that fall within the scope of this chapter are excluded from discussion on the grounds that they are arguably not central to a sustainable transport theme. Examples include policy debates around the humane treatment of livestock used in animal-drawn transport, the universal access

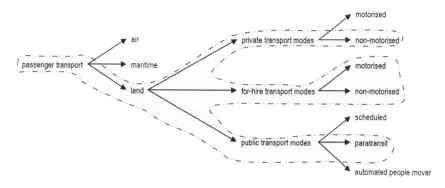

Figure 18.1 Passenger transport mode categorisation.

of transport systems vis-à-vis special services for persons with movement disabilities, and the exploitation of informally employed labour within the paratransit sector. Other policy issues that might be expected to feature in a paper with this particular focus are not included on the grounds that contemporary published literature relating to the subject was not found in the review undertaken. Examples of such policy debates include the accommodation of recreational non-motorised modes (e.g. rollerblades and skateboards) in public spaces, and appropriate regulatory frameworks governing the hiring of vehicles (e.g. chartered buses, hire cars, hire bicycles). There is also likely to be an (South) Afrocentrism in the interpretation of what policy debates and challenges are key, and what responses are appropriate. A further limitation is that facilitation and management of paratransit and for-hire transport are not areas in which I have conducted extensive research previously, so the depth of understanding of policy issues is likely to be uneven.

Outline of sections

The chapter is divided into five sections. The next section discusses key policy debates relating to private non-motorised modes. The section 'For-hire motorised and non-motorised modes' discusses key policy debates relating to for-hire modes and 'Paratransit modes' discusses debates relating to paratransit modes. 'Conclusion: The way forward' offers a brief and tentative exploration of key considerations in moving forward.

PRIVATE NON-MOTORISED MODES

Definition

Private non-motorised modes can be categorised into two groups on the basis of mechanisation. Non-mechanised modes take the form of walking and animal riding. Mechanised modes take the form of animal-drawn carts, human-drawn carts (e.g. trolleys, perambulators), cycles, mechanisms to assist persons with physical movement disabilities (e.g. wheelchairs, walking frames), and various recreational devices (e.g. rollerblades, skateboards, roller-skates, push scooters) (Figure 18.2).

Current policy debates and challenges

Looking beyond the proselytising nature of much of the NMT literature, the key policy debates and challenges relating to NMT modes are identified as:

- investigating the implications that a policy recognition of walking as a travel mode has for the development of appropriate analytical tools

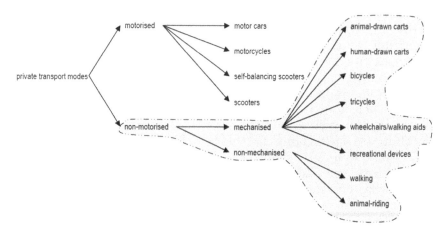

Figure 18.2 Private transport mode typology.

- unravelling the implications that NMT prioritisation policies have for appropriate transport system planning and design practices
- establishing institutional and implementational frameworks through which NMT improvements can be made systematically and proactively

Analysing walking as a travel mode

The modal share of walking in large cities of the 'developing world' is high. Recent data on main mode walking trips from Africa and Asia indicate that between 25% and 50% of main mode trips are on foot (Behrens 2004; Gleave et al. 2005; Gwilliam 2002; Tiwari 2002). Walking mode share is even higher in medium and smaller cities – in the region of 60–70% (Gwilliam 2002). When trip segments on either side of public transport main mode trips are taken into account, the role of walking becomes even more apparent.

Despite this considerable, if not the majority, share of modal split, walking has been ignored or under-recognised in transport policy and planning. Gwilliam (2002) argues that the welfare of pedestrians, and particularly the welfare of mobility-impaired pedestrians, has frequently been sacrificed in planning to increase the speed of vehicle flow

It is only recently that some transport policies have explicitly recognised the role of walking and dealt with it, in policy rhetoric at least, as a travel mode in its own right (as opposed to simply as an occasional supplement of other modes). The full implications of this recognition, however, remain incompletely explored, and methods of analysing walking patterns and pedestrian behaviour to inform planning and management practices remain underdeveloped. To be given equal priority in the design and management of transport infrastructure, data on all aspects of walking patterns and pedestrian behaviour will be required.

Conventional methods of travel analysis are typically limited in scope to motorised modes, commutes and peaks, a limitation that, as demonstrated by Behrens (2004) in the South African context, creates omissions and has in all likelihood distorted widely held perceptions of travel needs and patterns. More specifically, a perception can be created that non-motorised, non-work and off-peak trips are quantitatively and qualitatively significantly less important than motorised, work and peak trips. In particular, as a result of often being excluded or underestimated in past travel surveys, the importance of walking trips – in terms of their roles in satisfying travel needs and in analysing road safety problems – has not been fully understood. At best, these limitations on the scope of travel analysis have introduced a routine bias in the way in which the urban transportation problem has been framed and has skewed transport planning resources away from local and non-motorised network issues; and, at worst, because the greatest misrepresentation of travel needs occurs among lower-income households, they have led to neglect in the planning and design of infrastructure improvement for the poor and vulnerable road users.

A tiered hierarchy of methods to analyse walking patterns and pedestrian behaviour is needed, enabling analysis ranging from why, where and when pedestrians walk, to route selection and crossing behaviour, to infrastructure adequacy and site auditing. Some of these tiers of analytical method are in place, particularly methods relating to the capacity analysis of walkways and queuing areas, which have a long history of development dating back to the 1970s, but the development of other tiers is still in its infancy and a consensus on what represents 'good' or 'best' practice has yet to emerge. In many cities, basic questions of how far pedestrians walk, the routes they select, the purpose of their trips, and their origins and destinations cannot be answered.

Prioritising NMT in planning and design practice

Gwilliam (2002) argues that an explicit policy focus on NMT is necessary to redress a vicious cycle in past policies that biased the planning and management of urban transport systems in favour of motor vehicle users, at the expense of NMT users. A failure to recognise NMT, and invest in NMT infrastructure, led to NMT becoming less safe, less convenient and less attractive, which reduced NMT use, which in turn led to policy makers paying less attention to NMT modes and regarding NMT infrastructure investment as potentially wasteful. Many transport policies now, however, explicitly prioritise NMT and public transport modes over private transport on sustainability and equity grounds. This necessitates that the needs of NMT and public transport users are being considered and accommodated as a priority, from the outset of transport infrastructure planning and design processes, rather than considered after construction, when it inevitably becomes apparent that some form of retrofit is required to address design inadequacy. The full implications this has for how practices relating to the planning and design of transport systems should be changed have yet to be fully unravelled, and the accuracies of behavioural assumptions underlying numerous planning and design conventions, in 'developing world' cities at least, have yet to be fully interrogated. Good examples of such conventions are the spacing of freeways and arterials in cities and their associated (often implicit) underlying assumptions of walking trip distances and origin-destination pairs (see Behrens 2005 for an elaboration).

Even in contexts where NMT continues to receive little policy recognition and prioritisation, it can be argued that to address the needs of motorised modes it is necessary to also address NMT needs. Tiwari (2002), writing on the Indian context, observes that failure to meet NMT needs leads to suboptimal conditions for all modes of transport. She argues that if infrastructure does not meet the needs of NMT users, they are forced to use facilities that are not designed for them, and all users are forced to operate under suboptimal conditions. In her view it is possible to redesign street systems to meet the needs of diverse modes. This requires not only altering road geometry design and traffic management practices but also legitimising and accommodating the for-hire NMT services provided by the informal sector (an issue picked up in subsequent sections).

How practices relating to the planning and design of transport systems should be changed to adequately accommodate, or even prioritise, NMT modes is the subject of debate on some quite fundamental issues. One such debate relates to whether street networks should be configured to be discontinuous or multidirectional (i.e. grid-like). A considerable body of research has emerged in the United States on whether or not pedestrian- and cycle-friendly so-called 'new urbanist' street patterns have the effect of reducing vehicle kilometres travelled and increasing walking and bicycle use (see Boarnet and Crane 2001 for an overview of this literature). Another emerging debate surrounds so-called 'naked streets', and whether the safety interests of NMT users are best served by the provision of high levels of signage and road marking to control and regulate road user behaviour, or by stripping streets of this signage and marking and relying upon the uncertainty created to prompt more careful and negotiated road use (see, for instance, Firth 2011). Further debates relate to conditions under which NMT modes require rights-of-way separate from motorised traffic, and, when these are provided, whether or not pedestrians, cyclists and human-drawn carts should share the same path infrastructure or be separated onto different paths with associated costs (see, for instance, Servaas 2000).

Establishing systematic and proactive implementational frameworks

In the absence of adequate policy recognition of NMT, with notable exceptions, past initiatives to improve NMT infrastructure have typically occurred through opportunistic add-on projects, supportive of larger road or public transport improvement schemes. There would appear to be a reasonable amount of consensus in the NMT literature that this is inadequate, and an important policy challenge in both the 'developed' and 'developing' worlds is to establish clearly defined and appropriately funded institutional and implementational frameworks through which NMT improvements can be made in a systematic (or area-wide) and pro-active manner. With respect specifically to bicycles, Hook (2002) identi-fies the Dutch national bicycle master plan, and the city bicycle master plan implemented in Delft, as exemplary. Ideally, such plans should be inclusive of all NMT modes and closely integrated with strategies to improve road and public transport systems, as well as land-use manage-ment plans.

Gwilliam (2002) argues that the components of NMT strategies should include: (a) clear provision for the rights as well as responsibilities of pedes-trians and bicyclists in traffic law; (b) formulation of a national strategy for NMT as a facilitating framework for local plans; (c) explicit formulation of a local plan for NMT as part of the planning procedures of municipal authorities; (d) provision of separate infrastructure where appropriate; and (e) incorporation of standards of provision for bicyclists and pedestrians in new road infrastructure design.

FOR-HIRE MOTORISED AND NON-MOTORISED MODES

Definition

For-hire modes can be categorised into two groups on the basis of propulsion. Motorised for-hire modes take the form of hail taxicabs, 'demand responsive transport', motorcycle taxis, chartered buses and hired cars. Taxicabs take two main forms: conventional metered taxis and shared-ride taxis, in which two or more passengers with a common destination share the taxi fare. 'Demand responsive transport' (DRT) is an emerging form of flexible service that utilises communication tech-nology and intelligent transport systems (ITS) to facilitate short-term operator planning of destinations, routes and schedules in response to passenger demand. Motorcycle taxis take the form of two-wheelers prev-alent in West Africa (known as *okada* in Nigeria, *zemidjan* in Benin and *mototaxi* in Cameroon) and two-seater three-wheelers in South-East Asia (known as *tuk-tuk* in Thailand, *bajaj* in Indonesia and *selam* in Vietnam). Non-motorised for-hire modes take the form of human-drawn rickshaws (known as *rikisha* in Japan and Singapore), cycle taxis and hire bicycles (typically at tourist destinations or at public transport inter-changes). Cycle taxis take the form of bicycle taxis (known as *boda boda* in Uganda), tricycle taxis (known as *cyclo* in Cambodia and *velotaxi* in Germany) and quadcycles (Figure 18.3).

Current policy debates and challenges

The key policy debates and challenges relating to for-hire modes are identified as:

- the appropriate accommodation and management of cycle taxis and rickshaws
- the appropriate regulation and support of DRT services
- the appropriate regulation of hail taxicabs

Accommodating and managing of cycle taxis and rickshaws

In the past policymakers in some 'developing world' cities have banned or limited the operation of cycle taxis and rickshaws on the grounds that they are disruptive to the flow of motorised traffic, and that they reduce the potential ridership of subsidised public transport services. There appears to be a growing consensus in the literature, however, that these arguments are flawed, and that cycle taxis and rickshaws should be legitimised and accommodated (see, for instance, Cervero 1992; Gleave et al. 2005; Howe and Maunder 2004; Tiwari 2002). It is argued that facilitating private vehicle flow should not be the overriding policy concern for various poverty alleviation and equity reasons, and that, because the passenger capacity of for-hire modes is limited to a couple of passengers, these modes do not typically pose a viability threat to subsidised mass public transport carriers.

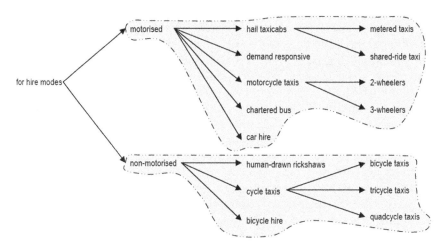

Figure 18.3 For-hire transport mode typology.

Problematically, in policy debates, for-hire modes have typically been lumped into the same category as paratransit,[1] but the arguments for and against regulation (to be discussed in greater depth later) are different in some respects.

Facilitating ease of market entry through deregulation is not as contentious an issue as it is in the case of paratransit services, which have far greater passenger capacities, although, as in the case of paratransit deregulation arguments, some form of quality or safety regulation remains desirable in the public interest. Further, as illustrated by Cervero (1992) in Southeast Asia, Gleave et al. 2005 in Cameroon and Howe and Maunder (2004) in Uganda, in some contexts at least, these modes operate in market niches and in contexts in which public transport and paratransit services would be unable to survive. More specifically, they operate in areas with low passenger demand profiles, or with decayed road infrastructure, which are suited to low capacity, two- or three-wheeled vehicles.

Regulating and supporting 'demand responsive transport' services

In much the same way that for-hire modes in the 'developing world' have filled gaps in the passenger market not catered for by commercial public transport and (regulated) paratransit services, 'demand responsive transport' in the 'developed world' has emerged in recent years to meet the needs of a section of the passenger market not met by conventional bus services (due to dispersed activity centres and dispersed patterns of demand), or by 'dial-a-ride' services for disabled and elderly users (due to the specialised and inflexible nature of these services) (see Brake and Nelson 2007; Enoch et al. 2004; Mageean and Nelson 2003; Palmer et al. 2004 for discussion on the emergence of DRT in Europe and North America). DRT is characterised by route flexibility, flexibility of booking methods and pre-booking regimes, and the use of ITS technologies, primarily in the form of global positioning systems, digital mobile radios, cellular telephones and dispatching and call reservation software.

Brake and Nelson (2007) argue that these recent innovative DRT services tend to operate independently of each other, leading to overlap and gaps. They argue that to address this problem, future DRT services will need wider area network planning, greater cooperation between service providers and improved understanding of passenger requirements. Enoch et al. (2004) observe that existing research on DRT has tended to focus on the technical means of delivery (i.e. types of vehicle, application of technology,

---

[1] The distinction between for-hire and paratransit modes made in this chapter is that the trip destination of the former is determined by the passenger(s), while, in the latter, the destination and route is determined by the operator (perhaps with occasional detours if passengers request and operating permits allow).

etc.), not on cooperation issues and associated regulatory, fiscal and institutional barriers at government, municipal and operator levels. They suggest these issues are as important as the more technical delivery issues.

Debate exists on how more coordinated DRT services, and a better interface with scheduled public transport services, might be achieved. Mageean and Nelson (2003) argue that the more regulated the environment, the less conflict there is between DRT and scheduled public transport modes. They contend that operators with a monopoly can plan services as they see fit, which should lead to a service without duplication, gaps and a fear of losing customers to competitors. They further argue that DRT tariffs are unlikely to cover the cost of the service in any market, and therefore need to be regarded by the public authority as fulfilling a societal obligation, and by the operator as a means of increasing passenger numbers. They suggest the provision of subsidies to cover operating cost deficits are most likely in regulated, as opposed to unregulated, environments. Brake and Nelson (2007) suggest that when operating in a deregulated environment, greater cooperation between service providers, and therefore more coordinated DRT services, could be achieved through partnerships.

Regulating taxicab services

Economists have advanced arguments for and against government regulation of economies through observations of the effects of regulation and deregulation policies on the taxicab industry. Regulation has typically been criticised on the grounds that it leads to a misallocation of resources, stifled technological advancement and lower productivity (Marell and Westin 2002). It is also argued to be expensive for authorities to administer and enforce.

In response to such criticisms, deregulation policies (principally in the form of the elimination of entry and price controls) have been applied in some (regulated) markets perceived to be inefficient and poorly performing. Marell and Westin (2002) summarise the goal of deregulation as the achievement of intensive competition by promoting new operator entry into the market. It is argued that if an industry consists of a few actors operating within oligopolistic conditions, these actors are interdependent and become aware that conflicting action worsens the situation for all, and therefore become competitively passive. Three beneficial economic effects are argued to follow from healthy competition: it is argued to lead to effective resource allocation, to result in lower prices and to stimulate innovative behaviour and development.

Some authors have attempted to observe whether these benefits have followed the deregulation of the taxicab industries. Marell and Westin (2002), for instance, examined the effects of taxicab deregulation in rural areas of Sweden in the 1990s. They found that, although not all expected benefits were fulfilled, some positive effects concerning competition and productivity were achieved. Gaunt (1995) examined the deregulation of the taxicab industry in the urban areas of New Zealand (with populations of less than 100,000) in 1989. Similarly, he found that a deregulated taxi market is competitive.

In contrast to these findings, Cairns and Liston-Heyes (1996) argue that experimentation with taxicab deregulation policies in some large American cities in the 1980s led to severe problems, and eventually to their abandonment. They suggest that deregulation of fares and entry may not be optimal. They argue that modelling the cruising taxicab industry as competitive would imply large numbers of firms facing large numbers of customers at a given instant at a given place, but suggest that these conditions of competition do not hold even approximately in the industry. They argue, for instance, that there are times when there are few customers and large numbers of taxicabs (e.g. at a taxi stand), in which the taxicab driver may prefer a fixed fare over facing a disadvantageous bargaining situation. At other times the inverse could apply. They argue price regulation is necessary, and entry regulation may be useful in some situations.

PARATRANSIT MODES

Definition

Paratransit modes take five main forms. Route shuttles provide unscheduled services on a route with a defined origin and destination (e.g. between a city centre and an airport). Motor tricycle taxis transporting six or more passengers are a higher-capacity version of for-hire two-seaters. Larger capacity carriers take the form of minibuses with around 15 seats (known as *dala dala* in Tanzania, *colectivo* in Argentina and Mexico, *público* in Puerto Rico and *bemos* in Indonesia), midibuses with around 25 seats (known as *jeepney* in the Philippines, *matatu* in Kenya and *trotro* in Ghana) and buses with around 50 seats. While some paratransit vehicles might be regarded as unroadworthy and poorly maintained, others (typically owner-driver operated) are flamboyantly adorned – with the remodelled US Army *jeepneys* of Manila, the *matatus* of Nairobi and the buses of Karachi providing some of the better-known examples of vehicle retrofit and decoration (Figure 18.4).

Current policy debates and challenges

The key policy debates and challenges relating to paratransit modes are identified as:

- the appropriate regulation of paratransit services
- the appropriate formalisation and recapitalisation of paratransit services, and their integration into (or coordination with) scheduled fixed-route public transport systems.

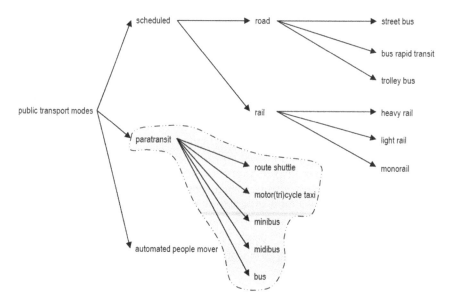

Figure 18.4 Public transport mode typology.

Regulating paratransit services

Much of the literature on paratransit deals with whether or not market entry should be regulated, typically through the issuing of operating permits or licences (typically relating to particular operating zones or corridors). Two main policy positions have been adopted in relation to this issue.

The first position argues that market entry should be deregulated to allow free competition between multiple operators, mediated by the 'invisible hand of the market', with public sector regulation restricted to public interest issues relating to vehicle roadworthiness and safe driving behaviour (see, for instance, Cervero's 1997 and 2001 arguments for deregulated paratransit in the United States to fill a market niche falling between fixed-route buses and exclusive-ride taxicabs). Cervero (2001) describes this as a policy of 'recognition', as opposed to 'regulation'. Appropriate policies of recognition involve the issuance and enforcement of rules and standards, mainly concerning areas of operations, safety, vehicle specifications and labour practices. In this view, compliance with minimum standards is the only legitimate form of entry restriction.

Some authors argue that, in the absence of effective public intervention, a degree of quality regulation, even if not adequate, can be achieved through self-regulation by operator cooperatives or 'route associations' (Cervero 2001; Golub 2003; Sohail et al. 2006).

Proponents of deregulation and free competition argue that it leads to reduced fares, reduced overall public expenditure, improved service levels, greater innovation and greater responsiveness to the needs of passengers. Paratransit services also have the allure of offering reasonably market responsive and penetrative services without the need for direct operator subsidisation. Cervero (1997, 2001) concedes, however, that paratransit competition is not without problems. In cities with high unemployment, unrestricted market entry can breed 'over-zealous competition' (e.g. drivers weave across lanes and cut each other off, stop in middle lanes to load customers, etc.) and oversupply, which can have negative effects on congestion and road safety. He further notes that 'hyper-competition' can lead to driver fatigue, vehicle overloading, traffic law violations, bald tyres, etc., which increase accident rates. He argues that such externality effects do not mean that governments should regulate paratransit entrepreneurs out of existence, but rather promote safety and fair competition, leaving matters of supply, service and price principally to the marketplace.

The second position argues that quantity regulation should be applied in addition to quality regulation. Iles (2005), for instance, argues that market entry deregulation can compromise route planning, which is best carried out by a large operator or by a regulatory authority, and can destroy established scheduled public transport operators, which are replaced by a disorganised service provided by a large number of small operating units. He argues that complete market entry deregulation may lead to over-concentration on lucrative routes and the removal of services from less lucrative routes or during off-peak periods (a practice known as 'cream skimming') and safety problems arising from competitive driving behaviour. He argues that some quantity regulation is therefore usually required, appropriate to circumstances. Sohail et al. (2006) argue that an important argument for retaining a degree of regulation in the provision of public transport is to ensure the provision of a network of services for the most vulnerable groups in society. They argue that deregulation can threaten the viability of transport services in marginal areas and restrict the accessibility of poor households to employment opportunities and social services.

Key to the argument for quantity regulation, or at least 'regulated competition', is that competition between paratransit and scheduled public transport services, particularly in 'developing world' countries, is not always 'fair'. Iles (2005) argues that the benefits of competition are only realised if all operators compete on equal terms. This is not the case when some operators fail to maintain their vehicles adequately, or drive dangerously in order to be the first to collect waiting passengers. In these unequal situations the operators who observe traffic rules, service their vehicles regularly and pay their drivers a reasonable wage are often forced out of business, to the detriment of the passenger. Golub (2003) describes this as 'ruinous competition'.

Formalising and integrating paratransit services with scheduled public transport

Further debate in the paratransit literature has dealt with how unroadworthy paratransit vehicles should be recapitalised, and whether paratransit services should be formalised and incorporated into scheduled public transport systems. Two main policy positions have been adopted in relation to this issue.

The first position – in line with the earlier arguments in favour of market entry deregulation and the limiting of public intervention to improving service quality and safety – argues that diversity in service offerings, and, almost inevitably therefore, competition with fixed-route scheduled public transport systems is desirable. Cervero (2001), for instance, argues that the urban passenger transport market benefits from an array of service and price options (i.e. an 'economy of scope'), rather than an economy of scale. He argues the inherent flexibility and profit motivations of competing and diverse paratransit services makes them market-responsive and more likely than public authorities to develop new services in response to changes in demand patterns (e.g. increased suburb-to-suburb commuting or off-peak travel). He suggests that in instances where paratransit vehicles compete directly with scheduled bus or train services, the policy objective should be simply to ensure that they do so fairly.[2]

The second position – drawing largely from the experiences of innovative 'bus rapid transit' (BRT) systems in some South American cities – argues that paratransit operators should essentially be formalised and coordinated with, if not assimilated into, contracted public transport operations. Wright (2004), for instance, identifies a spectrum of public transport services, ranging from what he regards as customer-unfriendly informal operations at one end to mass transit systems offering comfortable and high capacity services at the other, and within this he argues that most developing cities should be attempting to move towards the higher-quality end of the spectrum. He argues the lower-capacity and poorer-quality paratransit services often provide transit options for communities with few other choices. BRT is offered as a means to enter the higher-quality, higher-capacity end of the spectrum at a substantially reduced cost in comparison to rail services. He refers to a 'transit evolution' (perhaps with implicit modernisation undertones) from informal paratransit, to BRT systems with cleaner vehicles, sophisticated stations and fare collection systems and dedicated bus lanes.

Wright (2004) notes that past conventional wisdom has been that a wide diversity of public transport services in a city is beneficial, enabling different

---

[2] The notion of 'curb rights' has been proposed as a means of managing such competition between paratransit and scheduled bus services (see Klein et al. 1997). In essence, this involves assigning the rights to use particular street curbs at particular times to different operators, thereby managing the problem of paratransit operators scooping up passengers from a bus stop moments before the scheduled bus arrives (a practice known as 'interloping').

corridor operating conditions to be matched to an optimum mode. The current reality however, he argues, is often a plethora of services that are not integrated with each other and not understood by the majority of the population. He argues that BRT system innovations – which have enabled operating passenger capacities ranging from 4,000 to 40,000 passengers/hour/direction – have weakened the argument that diverse modes with fairly narrow bands of operational viability are required to match passenger demand across networks with varying volume and temporal profiles. He further argues that the cost of multiple mode technologies is high. Coordinating fare structures and distributing revenues within an integrated or coordinated system is complex and requires high-level managerial and administrative skills. Physical integration to facilitate passenger interchange can also be a challenge.

Iles (2005) argues there is, however, a role for smaller paratransit vehicles to serve as an effective complement to larger conventional buses – arguing the latter are generally a more effective form of public transport on main corridors, while the smaller vehicles are often more effective where demand is lower, or where the road configuration and surface quality makes the operation of larger vehicles impractical. Gwilliam (2000) observes that the establishment of appropriate pricing and charging devices and financing instruments to facilitate multimodal ticketing and inter-operator transfer of revenues represents a significant challenge, particularly when multiple small operators are involved. He argues that, as the role of the private sector increases, the more difficult the maintenance of multimodal ticketing systems becomes.

The establishment of, if not mutually acceptable then at least implementable, mechanisms through which to incorporate – very often fiercely, and sometimes violently,[3] competitive 'survivalist' – paratransit operators into improved and rationalised public transport systems presents another considerable policy challenge. In this respect, the experience of Bogotá possibly offers transferable lessons. An important aspect of the implementation of the BRT system in Bogotá (TransMilenio) was the scrapping of old paratransit vehicles in the contracting process. According to Ardila-Gómez (2004), for each new articulated bus, the concessionaires would have to purchase and destroy 2.7 old paratransit buses (increased to 7.7 buses in the second phase of system development). Through this mechanism, paratransit bus owners were compensated for the removal of their old buses. Other, better quality, paratransit buses were maintained in the system to serve feeder routes to the trunk high-capacity system. In addition, a condition for the new concessionaires was to have former paratransit bus owners among their shareholders. Another important aspect was the elimination of overzealous competition between bus operators arising from

[3] See, for instance, Dugard (2001) and Khosa (1992) for descriptions of so-called 'taxi violence' relating to route competition in South African cities.

paratransit drivers earning commission based on the number of trips and the number of passengers they transported. The system was replaced by one in which bus operators are paid per kilometre of service provided, and bus drivers earn a salary (Ardila-Gómez 2004). It is likely, however, that implementable agreements involving paratransit operators in other cities will only be achieved following highly context-specific negotiations around acceptable compensation, integration and business arrangements. Rather than offering a directly transferable solution, the Bogotá case illustrates the importance of institutional, as opposed to technological, arrangements in resolving problems associated with cut-throat driver competition and market overtrading.

CONCLUSION: THE WAY FORWARD

This chapter has reviewed literature dealing with modes of travel that are somewhat peripheral to the mainstream transport policy and planning literature. Relative to other fields, NMT, for-hire modes and paratransit are arguably under-researched, particularly in 'developing world' contexts. The nature of the policy debates and challenges extracted and synthesised illustrate that quite fundamental questions regarding appropriate future directions in policy and practice remain unresolved. This conclusion will explore, briefly and tentatively, what needs to be considered in moving forward.

With regard to NMT, a shift in mindset is required that acknowledges these forms of transportation, particularly walking, as travel modes in their own right, which require thorough analysis to facilitate the formulation of appropriate interventive strategies, and appropriate planning and implementation frameworks through which improvements can be made. With regard to for-hire transport, a clearer distinction from larger capacity paratransit carriers is required in the formulation of policies concerning market entry regulation, and an acknowledgement that these carriers serve particular market niches, and form important components of fragile survival strategies in 'developing world' contexts. It is apparent that different contexts require different regulatory regimes. With regard to paratransit modes, achieving an appropriate balance between quantity and quality regulation (including the regulation of intermodal competition when it occurs) and the effective integration of paratransit modes into rationalised public transport systems, are perhaps the most intractable and complex issues discussed in the chapter. It is likely that appropriate policies in this regard will be highly context-specific, dependent to a large extent upon the ongoing resources and managerial capacity of responsible public sector agencies, as well as the extent of entrenched interests among paratransit operators. These are highly variable, and there appears to be no easily transferable 'silver bullet'. Much of the recent BRT literature that advocates the incorporation of paratransit into improved and integrated public transport systems can be criticised (perhaps naively) for concentrating on technological questions surrounding vehicle selection, bus lane configuration, prepayment boarding systems, etc., at the expense of the far more complex and fraught, in many respects, 'soft' issues surrounding successful negotiation with fragmented informal paratransit operators with legitimate fears of losing their livelihoods and appropriate ongoing governance arrangements. Indeed, an overarching theme across all the modes discussed in this chapter is the absence of adequate attention in policy debates to 'softer' institutional and implementation issues. Progress in resolving these issues will prove central to the creation of more sustainable non-motorised, for-hire and paratransit modes.

REFERENCES

Ardila-Gómez, A., 2004. Transit planning in Curitiba and Bogota, roles in interaction, risk and change. PhD Dissertation in Urban and Transportation Planning, Department of Urban Studies and Planning, Massachusetts Institute of Technology, Boston, MA.

Behrens, R., 2004. Understanding travel needs of the poor: Towards improved travel analysis practices in South Africa, *Transport Reviews*, 24(3), 317–336.

Behrens, R., 2005. Accommodating walking as a travel mode in South African cities: Towards improved neighbourhood movement network design practices, *Planning Practice and Research*, 20(2), 163–182.

Boarnet, M. and Crane, R., 2001. *Travel by Design: The Influence of Urban Form on Travel*, Oxford University Press, New York.

Brake, J. and Nelson, J. D., 2007. A case study of flexible solutions to transport demand in a deregulated environment, *Journal of Transport Geography*, 15(4), 262–273.

Cairns, R. D. and Liston-Heyes, C., 1996. Competition and regulation in the taxi industry, *Journal of Public Economics*, 59(1), 1–15.

Cervero, R., 1992. Paratransit in Southeast Asia: A market response to poor roads?, Reprint No. 90, *Review of Urban and Regional Development Studies*, 3, 3–27.

Cervero, R., 1997. *Paratransit in America: Redefining Mass Transportation*, Praeger, Westport, CT.

Cervero, R., 2001. Informal Transit: Learning from the Developing World, *Access*, 18, 15–22.

Dugard, J., 2001. *From Low Intensity War to Mafia War: Taxi violence in South Africa (1987–2000)*, Violence and Transition Series, Volume 4, Centre for the Study of Violence and Reconciliation, Johannesburg, South Africa.

Enoch, M., Potter, S., Parkhurst, G. and Smith, M., 2004. *Intermode: Innovations in Demand Responsive Transport*, UK Department for Transport and Greater Manchester Passenger Transport Executive, London.

Firth, K., 2011. Removing traffic engineering control: The awkward truth, *Transport Engineering and Control*, February, 73–79.

Gaunt, C., 1995. The impact of taxi deregulation on small urban areas: Some New Zealand evidence, *Transport Policy*, 2(4), 257–262.

Gleave, G., Marsden, A., Powell, T., Coetze, S., Fletcher, G., Barrett, I. and Storer, D., 2005. A study of institutional, financial and regulatory frameworks of urban transport in large sub-Saharan African cities, Sub-Saharan Africa Transport Policy Program, Working Paper No. 82, The World Bank, Washington D.C.

Golub, A., 2003. Welfare Analysis of Informal Transit Services in Brazil and the Effects of Regulation. PhD Dissertation in Civil and Environmental Engineering, University of California, Berkeley.

Gwilliam, K., 2000. Public Transport in the Developing World: *Quo Vadis*? TWU Series Discussion Paper TWU-39, The World Bank, Washington D.C.

Gwilliam, K., 2002. *Cities on the Move: A World Bank Urban Transport Strategy Review*, The World Bank, Washington D.C.

Hook, W., 2002. Preserving and expanding the role of non-motorised transport, Division 44 Environment and Infrastructure Sector Project "Transport Policy Advice", Deutsche Gesellschaft für Technische Zusammenarbeit (GTZ), Eschborn, Germany.

Howe, J. and Maunder, D., 2004. Boda Boda: Lessons from East Africa's Growing NMT Industry, 10th World Conference on Transport Research, Istanbul, Turkey.

Iles, R., 2005. *Public Transport in Developing Countries*, Elsevier, Amsterdam, Netherlands.

Khosa, M. M., 1992. Routes, ranks and rebels: Feuding in the taxi revolution, *Journal of Southern African Studies*, 18(1), 232–251.

Klein, D., Moore, A. and Reja, B., 1997. *Curb Rights: A Foundation for Free Enterprise in Urban Transit*, Brookings Institution Press, Washington D.C.

Mageean, J. and Nelson, J. D., 2003. The evaluation of demand responsive transport services in Europe, *Journal of Transport Geography*, 11(4), 255–270.

Marell, A. and Westin, K., 2002. The effects of taxicab deregulation in rural areas of Sweden, *Journal of Transport Geography*, 10(2), 135–144.

Palmer, K., Dessouky, M. and Abdelmaguid, T., 2004. Impacts of management practices and advanced technologies on demand responsive transit systems, *Transportation Research Part A: Policy and Practice*, 38(7), 495–509.

Servaas, M., 2000. *The Significance of Non-Motorised Transport for Developing Countries: Strategies for Policy Development*, Interface for Cycling Expertise, Utrecht, Netherlands.

Sohail, M., Maunder, D. A. C. and Cavill, S., 2006. Effective regulation for sustainable public transport in developing countries, *Transport Policy*, 13(3), 177–190.

Tiwari, G., 2002. Urban transport priorities: Meeting the challenge of socio-economic diversity in cities, a case study of Delhi, India, *Cities*, 19(2), 95–103.

Wright, L., 2004. Planning guide: Bus rapid transit, Division 44 Environment and Infrastructure Sector Project "Transport Policy Advice", Deutsche Gesellschaft für Technische Zusammenarbeit (GTZ), Eschborn, Germany.

Chapter 19

Paratransit and non-motorized traffic as mainstream road users

Geetam Tiwari

CONTENTS

INTRODUCTION

The informal sector in Asian cities is an example of a self-organizing system that can provide us with useful insights to address future urban transport problems. This sector continues to be viewed as an "unwanted" sector in the city and hence formal plans of housing and transport do not have provisions for their needs. Yet this sector exists in every city and millions of people across Asia survive because of the innovations taking place in this self-organizing system.

The informal sector

Informality has been defined in many ways. It is outside what is official or legal or planned. It is certainly not a synonym for criminality, which is both outside the law and illegal. Squatter settlements all over the world are called informal settlements because they are not part of the official plan. Similarly the transport solutions and employment solutions that evolve because of necessity outside the formal plans are labeled as "informal".

The patterns of development in Asian megacities are amalgams of planned and organic self-organizing growth. Most of us see congestion, crowding, poverty and chaos as a ubiquitous phenomenon; however, we fail to recognize the human ingenuity for survival, the social cohesion and low street crime rates present in these cities. If Asian cities are allowed to follow the pattern of exclusion, we may also see a sharp rise in street crime. The important lesson from these trends is to recognize that an informal economy and marginal settlements are an integral part of cities.

Urban travel patterns

Urban travel in Asian cities is predominantly by walking, cycling and public transport (Table 19.1). The motorized and non-motorized paratransit services play an important role in providing affordable mobility to a large section of the citizens and form a large proportion of public transport trips. The variation in modal split between these three modes seem to have a relationship with city size and incidence of poverty (per capita income). Dhaka with 54% poverty incidence, the highest among Asian cities, is heavily dependent on walking and cycle rickshaws. Small- and medium-size cities (Kanpur, Ahmedabad) have lower incomes than megacities; therefore, dependence on cycle rickshaws and cycles is greater than in larger cities. Shanghai and Beijing, both megacities with high industrial growth, continue to depend on walking and cycling. In Shanghai, until about 1990, almost all travel was by foot, bicycle or bus. Cars, scooters and motorcycles were rare. Travel distances increased in the late 1980s and 1990s, not only because of income growth but also because of industry relocation. The movement of factories from the central city to the periphery created long commutes for many workers.

Table 19.1 Modal split of daily trips, selected cities in Asian countries

City (population in million)	Modal split (percentage of daily trips)						
	Walking	Cycles	Public transport	Two-wheelers	Car	Paratransit MTW[a]	CR[b]
Delhi (13)	14	24	33	13	11	1	
Mumbai (14)			88		7	5 (taxi)	
Kanpur (3)	34	18	12	23	0	4	9
Ahemdabad (5)	40	14	16	24	0	5	0
Beijing[c] (12)	14	54	24	3	5		
Shanghai[d] (13)	31	33	25	6	5		
Manila (10)			24 + 2 + 3		30	41	
Jakarta (11)	13						12
Dhaka[e] (14)	62	1	10	4	4	6	13
Bangkok (7)	16	8	30		46		

Notes: [a] Motorized three-wheeler taxi. [b] Cycle rickshaw. [c] Zhou He-Long and Deng Xin-dong, Cycling Promotion and Bicycle Theft, Report for Interface for Cycling Expertise, Utrecht, The Netherlands, 1996. [d] Ximing Lu, Xiaoyan Chen, and Xuncu Xu, *Urban Transport Planning and Urban Development*, Shanghai City Comprehensive Transportation Planning Institute, East China Polytechnic Publishing House, 1996. [e] Hoque M. and Jobir Bin Alam, Strategies for safer and sustainable urban transport in Bangladesh, proceedings. CODATU X, Lomé, 2002.

Similar policies were adopted in the mid-1990s in Delhi. Also, newly developed low-density areas were too far for bicycles and not profitable for buses, therefore scooters and small motorcycles became a popular mode of travel. Unlike Indian cities, where the scooter and motorcycle numbers are the fastest growing, the scooter and motorcycle numbers in Shanghai and other Chinese cities are declining because of new restrictions on the registration of new scooters and other vehicles with two-stroke engines.[1]

Delhi is showing declining trends in three-wheeler numbers because of restrictions on fuel and the age of public transport vehicles. All three-wheelers are required to run on compressed natural gas (CNG) and should be less than eight years old, by order of the court. Shanghai and Beijing are the only cities in this list where a planned bicycle infrastructure exists. This may explain the high share of bicycles in these cities despite their having higher incomes than the Indian cities.

Dependence on walking and cycling trips, despite the absence of an infrastructure for them, shows the presence of captive users of these modes. Since organized public transport has been found to be financially unviable in many of these cities, dependence on paratransit (cycle rickshaws and three-wheelers) has been increasing. However, due to restrictive policies (legal regulations and restrictions on movement) for these modes, the actual numbers of these vehicles present on the roads are often not included in official registration numbers.

In cities where the informal sector (housing and employment) is an integral part of the urban structure, though usually unplanned (Chinese cities were an exception until the 1980s), the two major trends that can be noticed are: (1) the predominance of walking, cycles and public transport, and (2) cities that lack formal public transport services are dominated by indigenously designed vehicles: three-wheelers in Indian cities, jeepneys in Manila, *tuk-tuks* in Bangkok, or cycle rickshaws in Dhaka and Kanpur. These are locally fabricated and require minimal training to operate. This is an example of an innovation to survive in the city, which is rewarding at an individual level, however, it is suboptimal at the societal level due to lack of safety and environmental concerns. It is also interesting to notice that, when restrictions are imposed on private vehicles such as motorcycles in Chinese cities, their numbers remain low. However, restrictions on paratransit modes in Delhi, Jakarta and Dhaka have resulted in erroneous official statistics and individual strategies to circumvent the law. Passenger *becaks* (cycle rickshaws) are converted to goods *becaks* in Jakarta at the time of inspection. Pedestrians and bicyclists continue to use the roads, despite the hostile environment, because they have no other choice.

---

[1] Zhou Hongchang, Daniel Sperling, *Transportation in Developing Countries: Greenhouse Gas Scenarios for Shanghai, China*, PEW Center on Global Climate Change, July 2001.

Innovations for survival

Urban residents have shown range of innovations to meet their basic mobility needs. Often this has meant violating the formal plans and vehicle safety standards.

Delhi: Squatter settlements have grown up close to places of work inside the city. Most work trips are by walking or by bicycle. Pedestrians and cyclists are present on all roads. Road hierarchy, using a textbook definition, does not exist. People from resettled housing some distance away from the city centers travel in goods vehicles, violating safety guidelines. The actual number of cycle rickshaws in the city is estimated to be three times the current official figures.

Jakarta: Large numbers of *becaks* continue to exist in the city despite an official ban. Passenger rickshaws are converted to goods rickshaws at the time of official checking. Each vehicle is shared by more than one driver to provide employment to more people. Three-wheeled taxis imported from India in 1975 continue to run on the road. Hawkers are found on the limited access expressway passing through the city.

Three-wheeled vehicles: Autos in Delhi, *bajaj* in Indonesia, *tuk-tuks* in Bangkok, *baby taxi* in Dhaka and larger vehicles like *jeepneys* in Manila, *tempos* in Indian cities and *bemo* in Surabaya are all locally manufactured. These vehicles do not meet safety or environmental standards, yet meet the mobility needs of the urban residents, which have not been met by the formal public transport system.

Streets are used according to local needs: Urban streets of Asian cities are used by pedestrians, non-motorized vehicles and motorized vehicles at the same time. However, the physical design is primarily influenced by the needs of motorized vehicles. In high-income countries, it has been possible to some extent to define the primary function of city streets, allowing a well-defined hierarchy of streets, along with suitable geometric designs, to be developed accordingly. Often the urban arterials have well-defined roles carrying relatively fast-moving car traffic. The road cross-section is designed to accommodate car traffic and parallel service lanes are provided to meet the needs of abutting land use, such as shops or the entrances and exits of residential or commercial areas. Well-defined shopping streets are often developed with wide pedestrian paths, street shopping and roadside restaurants. In contrast to this, the streets of Asian cities are often self-organized, multifunctional entities. The socioeconomic conditions of most cities influence the way streets are used. This is reflected in not only the mix of traffic that is present on the road, but also the other activities that are present along the road. Arterial streets are used by motorized vehicles as well as bicycles, cycle rickshaws and pedestrians. Street vendors, bicycle repair shops, etc., are common sights on arterial streets and near bus stops. Often, streets in Asian cities are not restricted to the movement of vehicles and people. The range of activities includes services required by the users and social activities that require safe public spaces.

Current policies

Currently, according to our understanding of the use of urban streets, the services required by street users are limited by the definition of legal, motorized modes. Yet, since the captive users have no choice, there is a mismatch between design and usage of the infrastructure.

Congestion solution: Restricting the numbers and movement of three-wheelers and rickshaws is promoted to relieve congestion. In Delhi rickshaws are not permitted on arterial roads. In the old Delhi area, which has narrow roads, the government recently ordered the removal of rickshaws and instead introduced small buses. The current policies regarding cycle rickshaws and other non-motorized vehicles are restrictive and are based on the false notion that efficient transport systems do not have any place for these vehicles. Traffic management experts and traffic police have proposed area and time restrictions on the movement of rickshaws in Delhi. The number of rickshaws that can be registered in the city is fixed by the government (99,000) in order to restrict the number of rickshaws. The registration procedure requires the owner to have a valid registration card. Rickshaws are allowed to be registered only during a stipulated time period twice a year. These restrictive policies must be viewed in the context of the present environment of a globalized economy, where the highest level of policy makers talk about reducing government controls in order to enable the free market economy to operate. Should these policies not apply to the operation of cycle rickshaws and other non-motorized vehicles as well?

Urban transport pollution solution: The public transport fleet in Delhi has been converted to run on CNG. While this has resulted in lower emissions of particulate matter, a large number of small buses have joined the fleet and government-run buses have been reduced in number. The formal public transport system has been replaced by an informal system where safety concerns have been compromised. The bus and three-wheeled taxi fleets have been reduced by 50%. The use of cars and two-wheelers continues to grow. Why is it that pollution reduction strategies do not discuss the needs of zero pollution vehicles – bicycles and rickshaws? Investment in the non-motorized vehicle infrastructure not only reduces pollution but also improve safety, because the largest proportion of road traffic fatalities involve pedestrians, bicyclists and motorcycle users. However, policymakers have not given any importance to such strategies. In the name of cleaning the environment, Delhi, Dhaka and Jakarta have imposed restrictions on cycle rickshaws.

The success story of improved environmental requirements has been the phasing out of lead. However, these technological fixes do not answer the mobility needs of captive users, the people without choices for whom restrictive laws have no meaning.

Future directions

The future solutions lie in resolving the conflicts that exists at several levels. (1) Planning for the informal sector in the city. (2) Meeting the needs of the captive users – pedestrians and cyclists prior to cars. (3) Ensuring safe accessibility and mobility in order to ensure a clean environment. This can happen when the problems of mobility, pollution and safety faced by different users pass through the filter, which is sensitive to their needs (Figure 19.1). Methodologies and engineering solutions can evolve once the new paradigm has been accepted.

An understanding of the symbiotic relationship existing between the informal and formal sectors is imperative for solutions. A poor understanding of this relationship leads to physical segregation, planned townships and slums being relocated away from the formal city. Poor families, who are also transport-poor, are then denied accessibility to economic opportunities and have the burden of reduced mobility imposed upon them.

Improvement in road capacity in Asian cities has meant reducing pedestrian and bicycle facilities, removing street vendors, restricting pedestrian movement and constructing grade-separated junctions. This leads to lonely streets and spaces that are of no interest to anyone, at the expense of substantial financial resources. Evidence from several Chinese and Indian cities shows that benefits to vehicular traffic are also short-lived. On the other hand, if the informal sector is considered an integral sector of the urban economy then the improvement in transportation involves changing the design criteria for urban arterials. Numbers and types of hawkers around a bus stop can be predicted on the basis of the number of waiting commuters. Cycle repair shops, cold drinks and snacks provided by street hawkers serve the same function that a well-designed service area serves along a

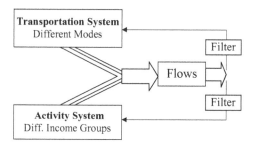

Figure 19.1 Transportation and activity system interaction.

highway. The difference is in the speed of users and therefore the difference in frequency and density of service providers. Thus, standards for geometric design of these roads would include spaces for hawkers and the presence of hawkers would not be seen as "encroachments".

Lessons from self-organizing systems for resolving conflicts

The perception of benefits and risks can be influenced by adopting the right methodology (priority to pedestrians and non-motorized vehicles).

A better understanding of the function of streets, space utilization, etc., will give us a different design criteria.

The perception of mobility benefits and risks can be influenced by realizing that:

- Access to work is the prime concern, not safe mobility.
- Benefits of increased speeds accrue to one subgroup, while the penalties are imposed on another.
- Self-organizing patterns of Asian cities provide us insights for creating inclusive urban spaces that welcome diversity and meet the contrasting needs of different social groups. This is central to the goal of building "A city for ALL."

Chapter 20

Politics of mobility and the science (?) of sustainability

Dinesh Mohan

CONTENTS

INTRODUCTION

The introduction to *Mobility 2030: Meeting the Challenges to Sustainability* includes the following statement (WBCSD, 2004):

One factor has continued to impress us throughout our assessment – the strength of people's desire for enhanced mobility. Mobility is almost universally acknowledged to be one of the most important prerequisites to achieving improved standards of living.

Followed by:

> First, transport services can be made more efficient, increasing the amount of economic growth supported by a given volume of transport services. Second, the level and composition of "induced" mobility demand can be channelled in ways that fulfill growing mobility needs but create fewer transport impacts. Third, the level of adverse economic and environmental impacts associated with any given level of transport activity can be greatly reduced – for example, through significant technology shifts. *Undertaking any or all of these can reduce – and perhaps eliminate totally – the threat that adverse economic and environmental impacts would be sufficiently great that transport services would be hindered in enabling economic growth. If this could be accomplished, mobility could be said to be sustainable.* (emphasis added)

This is the considered view of the sponsors of this report who include the major automobile and oil companies of the world. What is interesting in this worldview is that value-laden terms such as *"people's desire"* and *"standards of living"* are considered to be objective, neutral and universal. Even more interesting is the fact that they think that we can *"perhaps eliminate totally"* the adverse health and economic consequences of motorised transport through *"significant technological shifts"*. It appears that both the assumptions and conclusions of this report have a fairly weak factual base.

The main complications facing us in designing for sustainable urban transport are that the base conditions for promoting non-motorised modes and public transport are still not clear. In his book *Unsustainable Transport: City Transport in the New Century*, Bannister claims that we are yet to come to terms with the following issues (Bannister, 2005):

- City size and modal choices.
- City density and development – compact, sprawl, decentralised concentration. This is especially important for the expanding cities in low- and middle-income countries (LMIC), as most new development is taking place in the periphery.
- The location of business centres, the existence of shopping malls versus smaller stores.
- The impact of an intelligent transport system (ITS) on travel modes and frequency of travel.

The form of public transport: metro versus bus rapid transit.

It is interesting that Bannister does not list traffic safety as a major issue. In this chapter, we will try and deal with a few of these issues.

MODAL SHARES, PUBLIC TRANSIT AND SUSTAINABILITY

Delhi and Stockholm, two very different cities, are compared in order to understand travel modes and sustainability issues. Table 20.1 gives the demographic and population details of Delhi and Stockholm (Stockholm County Council, 2003, 2005; Planning Department, 2005). These data show that although Stockholm county is larger in area than Delhi, its city centre (Stockholm municipality) is much smaller. Delhi is a much larger city in terms of population, with an average density about twice that of Stockholm municipality. The average income in Stockholm is about 25 times that of Delhi in absolute terms and 4 times greater in terms of purchasing power parity.

Table 20.2 shows the vehicle ownership details for Stockholm and Delhi (Stockholm County Council, 2003, 2005; Planning Department, 2005). It is interesting to note that the per person car ownership rate in Stockholm is 7.5 times more than that in Delhi, which is about double the purchasing power parity (PPP) income ratio but one-third the absolute income ratio. One-fifth of the number of families owns a car compared to that in Stockholm. This ratio is low because the family size in Delhi is 2.5 times greater than in Stockholm. However, when one includes motorcycle ownership in the two cities the gap in vehicle ownership per family decreases

Table 20.1   Demographic and population details of Delhi and Stockholm

	Population	Area sq km	Population density pers./sq km	Household size pers./hshd	Gross income per capita Euro	Gross income per capita PPP Euro
Delhi	14,400,000	1,483	9,710	5	950	5,300
Stockholm County	1,850,467	6,490	285	2	24,246	20,400
City of Stockholm	758,148	187	4,048	2	22,277	18,700

Table 20.2   Vehicle ownership details in Stockholm and Delhi

City	Cars per 1,000 persons	Motorcycles per 1,000 persons	Total personal vehicles per 1,000 persons	Families with one or more cars (%)	Percentage of families with one or more personal vehicles (%)
Delhi	50[a]	74[a]	124	13	42
Stockholm County	373	20	393	64	66[b]

Notes: [a] Number based on estimated number of vehicles on the road in Delhi in 2002. From *Report of the Expert Committee on Auto Fuel Policy*. 1–98. 2002. New Delhi, CSIR, Government of India.
[b] Estimate based on motorcycle ownership.

significantly – per capita ownership in Stockholm is 3 times that in Delhi and the percentage of families owning a personal vehicle only 1.5 times in spite of the huge gap in per capita incomes. This factor is bound to have a significant impact on travel patterns and modal shares in transport use.

Table 20.3 shows the transport modal shares in the two cities. The proportion of personal motor vehicle use (including motorcycles) in Stockholm is 2.4 times that in Delhi, which ratio is in between personal and family ownership rates in the two cities. The proportion of public transport trips in Delhi are just 15% less than in Stockholm, but non-motorised modes 10 times greater. In addition, the availability of taxis per capita is double in Delhi. Just looking at these figures would indicate that, by any objective criteria, the transportation situation in Delhi is what transport planners would term as more sustainable and desirable compared with that of Stockholm!

However, this is not borne out by the road safety and air pollution data shown in Table 20.4 (Ojha, 2002; Tegner, 2003) and Table 20.5 (NEERI, 2003; World Bank, 2006). These tables indicate that threat to life due to road traffic injuries and pollution in Delhi is much greater than that in Stockholm. This shows the complexity in planning transport for a sustainable future. It is clear that having the right modal mix and low use of personal transport is a necessary but not a sufficient condition for the provision of clean air, safe roads and optimal access conditions in cities.

It is also possible that, if public transport becomes more expensive, due to the use of more costly clean technologies, some road users may shift to sharing old cars or using motorcycles, thus increasing pollution. These data raise even more important issues regarding the promotion of public transport. Compared to Stockholm, Delhi has a lower proportion of public transport users but a higher proportion of walking and bicycle trips. *With the*

Table 20.3   Transport modal shares in Delhi and Stockholm

	Modal share (%)				Taxis per 1,000 persons
	Car	Motorcycle	Public transport	Bicycling and walking	
Delhi	8	14	38	40	6[a]
Stockholm County	52	n/a	45	4	3

Note: [a] Including three-wheeled scooter taxis.

Table 20.4   Traffic safety data for Stockholm and Delhi 2002

	Fatalities per million population	Fatalities by road user type (%)				
		Car	Motorcycle	Bicycle	Pedestrian	Other
Delhi	118	3	21	10	53	13
Stockholm County	33	66	4	17	13	0

Table 20.5   Air pollution data for Stockholm (2000) and Delhi 2002

City	Annual mean, 24 h average values (μg/m³)			
	SPM*a*	RPM*b*	SO₂	NO₂
Delhi	476	179	9	48
Stockholm County		15	3	20

ᵃ Suspended particulate matter
ᵇ Respirable particulate matter

rise in incomes and the promotion of public transport, these non-polluting trips (walking and bicycle) can shift to polluting modes (public transport). Therefore, merely the promotion of public transport may not give us the more sustainable future we want in LMICs. This is also complicated by some changes in technologies and modes of living over the past century.

HISTORY AND POLITICS OF URBAN TRANSPORT

Professionals, poverty, city size, transport and the demands of the rich

Higher education and trade obviously have a reasonable amount to do with the sizes of cities and the degree of urbanisation. The more "educated" we are, the larger the pool of resources we need for both work and human contact. Therefore, a large city becomes essential for a reasonable section of the population for finding "optimal" employment and friends. The inverse of the same issue is that trade and industry need a large pool from which to select employees. This forces LMIC cities to become larger than cities in high-income countries (HICs). This is because in the LMICs there is a much larger proportion of poor people compared to in the HICs. So, the same number of professionals in an LMIC will coexist with a much larger number of poorer residents than that in an LIC. For the foreseeable future, this will make LMIC cities much larger than the "mature" cities of Europe. *The existence of a large number of low income people pursuing informal trade and income generating activities places different political pressures on the rulers, increases demand for low cost mobility and short distance access to jobs and trade.*

This is offset by the middle- and upper-classes wanting to live away from the poor and form gated communities at the periphery of the city. These developments set up a powerful political demand, aided and abetted by contractors and consultants, to provide infrastructure. The upper-middle-class of the post-colonial nations mainly have the United States as a model for the good life. Especially those nations that have English as an elite language (like ex-British colonies). All Asian, African and South American cities are more influenced by the US than by any other society. For example, American town planners were in Delhi helping us plan our cities in the

1950s (courtesy of the Ford Foundation). So, all these cities have tried hard zoning, with broad avenues and with highways running through them. If it hasn't happened, it is due to inefficiency and lack of money!

However, cities with somewhat soft governments have developed more organically. Illegal settlements, illegal trade and the informal sector have made our cities develop more logically and in a more sustainable manner. The poor live closer to work, spend more time at home – with better family life and stability. They don't need public transport in as large a quantity as those living in settlements in South Africa. So, cities in countries such as India are largely more human scale, sustainable and liveable, at least if you are middle class. *The politics of sustainable transport will revolve around the power which the poorer sections of the population can exert on decision-making. Wherever the lower income groups are able to get themselves heard, we are more likely to have more sustainable cities as they will need facilities for walking, bicycling and public transport closer to their places of work and for shopping and leisure activities around their homes. This will influence what sustainable cities will look like in the future. The upper class is unlikely to do it willingly.*

Changes in the urban transport scenario in the last century – differences between HICs and LMICs

Politics, ideology and cities

Most large, HIC cities grew to their present size between 1850 and 1950. This was the period in which Marxism, or socialism, were accepted by intellectuals, a large proportion of politicians and the working classes as ideologies of choice. This influenced the location of schools, people's living arrangements and the provision of public transport as a public responsibility. However, city growth in most LMIC cities has not been strongly influenced by these ideologies in a comprehensive manner over the past five decades, thus favouring the provision of facilities for the upper class and for cars. This is likely to continue for some time.

CBDs in HICs and sprawl in LMICs

Public transport facilities in the "mature" cities of HICs were initiated in the second half of the nineteenth century. This was long before the car and so everyone had to use public transport or walk/bicycle to get to work, including the middle classes. Car ownership started to increase in the 1920s, but most families did not own a car until the middle of the twentieth century. By then, the essential land-use and transportation patterns of large cities were well established with large central business districts (CBDs). This encouraged the building of high-capacity grade-separated metro systems, and, in turn, the transport system encouraged the densification of CBDs as large numbers of people could be transported to the centre of the city. The non-availability of the car to the middle classes decided the widespread use of public transport and city structure. Car ownership increased much faster in the US, and so many cities did not have the political pressure to provide for public transport.

Most LMIC cities have expanded since 1960. In India and many other countries, large cities have planned for multiple business districts. In addition, in the second half of the twentieth century most families did not own a personal vehicle and so all leisure activity took place within short distances from the home. In the past two decades, motorcycle ownership has increased substantially in many Asian cities. As a result, about 50% of Delhi's families own a car or a motorcycle at a very low per capita income level of about USD 1,200 per year. Such high levels of private vehicle ownership did not happen until incomes were much higher in HIC cities. Therefore, the *high level of ownership of motorcycles, the non-availability of funds to build expensive grade-separated metro systems and official plans encouraging multi-nodal business activity in a city have resulted in the absence of dense high population CBDs and city forms that encourage "sprawl" in the form of relatively dense cities within cities. This further obviates the need for very high-capacity public transport systems.*

Car technology changes and declining demand for public transportation

Most middle-class families did not own air-conditioned cars with stereo systems in HICs before 1970. The cars were noisy and occupants were exposed to traffic fumes as windows had to be kept open. Under such conditions, the train was much more comfortable. This created the conditions in which there would be a political demand for metro systems that came from the middle classes and could not be ignored. On the other hand, brand new, quiet, stereo-equipped, air-conditioned cars are being sold in LMICs now at prices as low as USD 5,000–6,000 and used ones for half this price. This has made it possible for the middle-class first-time car owner to travel in cars with comfort levels Europeans had not experienced until the late twentieth century. Air-conditioned, comfortable, safe and quiet travel in cars with music in hot and tropical climates cannot be matched by public transport. Owners of such vehicles would rather brave congestion than brave the climate on access trips and the jostling in public transport. *If public transport has to be made more appealing, it has to come closer to home, reduce walking distances and be very predictable. These conditions would favour high-density networks, lower-capacity, surface transport systems (to reduce walking distances) with predictable arrival and departure times aided by ITS information systems.*

Motorised two-wheelers and public transport

Wide ownership of motorised two-wheelers (MTWs) has never been experienced by HIC cities. This is a new phenomenon, especially in Asia. The efficiency of MTWs – ease of parking, high manoeuvrability, ease of overtaking in congested traffic, same speeds as cars and low operating costs – make them very popular, despite MTW travel being very hazardous. The availability of MTWs has further reduced the middle-class demand for public transport. In addition, it has pegged the fare levels that can be charged by public transport operators. It appears that public transport cannot attract the road users who can afford an MTW unless the fare is less than the marginal cost of using MTW. At current prices this amounts to less than USD 0.02 per km. *It is difficult to operate any clean public transport system at this fare, certainly not elevated or underground systems. The only option available is to design very cost-efficient surface transport systems.*

The above discussion shows that the LMIC cities in the twenty-first century are growing under very different conditions from those for LMICs in the first half of the twentieth century. The political and ideological forces combined with changes in technologies will make it difficult to provide efficient transport systems in the old manner. It will also be very difficult to move away from multi-nodal city structures with future job opportunities developing on the periphery. This challenges us to design surface public transport systems that are efficient, are medium-capacity, have dense networks, are ITS-enabled, and can be fully integrated with taxi systems.

CONCERNS BEYOND TAILPIPE EMISSIONS

Socialisation

Health and environmental issues that go beyond those highlighted up to now will have to be given more importance to get the support of a wider section of the population to control car-oriented road and city development policies. The World Health Organization has highlighted the broader health issues concerning the health of children and the elderly, obesity, socialisation and learning issues (2000). Table 20.6 and Figure 20.1 reproduced from this report show the negative effects of heavy traffic and wide streets. These are not widely known among the policy makers or as policy issues, though citizens themselves have negative reactions to these developments. Reports have even quantified the negative effects of noise on rental values. A lower volume of pedestrians and bicyclists removes social control from the streets, leading to more crime, further discouraging people from walking. Narrower arterial streets with designs obviously favourable for the pedestrian have to be put centre stage. Such criteria could include mandatory tree cover, convenient toilets, places for kiosks and street vendors, etc.

Table 20.6   Road traffic and networks of social support

Traffic levels	Contacts living on the same street	
	Friends	Acquaintances
Light traffic (200 vehicles at peak hour)	3.0	6.3
Moderate traffic (550 vehicles at peak hour)	1.3	4.1
Heavy traffic (1,900 vehicles at peak hour)	0.9	3.1

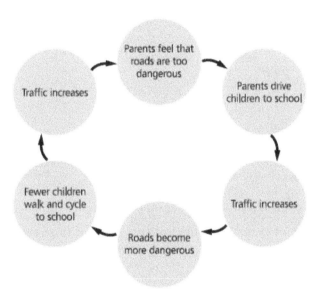

Figure 20.1   The effects of ever-increasing traffic on children's freedom of movement.

Water use

Water use is not normally considered in debates on sustainability. However, most cities in India and other LMICs suffer from a perennial drinking water shortage. A car uses water in many forms including its regular washing. Very often this is from the municipal drinking water supply available to rich households but not to the poor. The water use issue seems to be more serious. Figure 20.2 shows the calculated values for lifetime water use in Finland by different modes of passenger transport (Saari et al., 2007). The highest value of water use per person km is 28 kg for cars using connecting roads. Value for bicycle use is high as, in Finland, bicycle use is very low for an extensive network. For cities likely Delhi it is likely to be 1/10 or less than that shown in Figure 20.2. It would be useful to add water use issues to the list of adverse effects of personal mechanised transport.

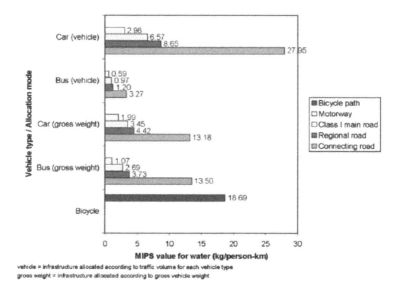

Figure 20.2 Material input per service unit values for water consumption in the case of passenger transport.

Hidden increases in energy consumption

Ailments, lack of exercise and obesity would also increase energy use, and ultimately greenhouse gases, in the following activities:

• Energy used by hospitals and in traveling to a doctor
• High-tech exercise and health centres
• Manufacture of health aids and drugs

CHOICES AVAILABLE AND NEED FOR KNOWLEDGE

Public transport

There are 36 cities in India that had a population of more than 1 million in 2001 (Census, 2001). This number is likely to increase significantly in the next two decades. Three of these have populations in excess of 10 million and another two may join them. Another five to eight of these may reach populations of 5 to 10 million. Therefore, it is necessary to plan for providing efficient access and mobility to citizens of a large number of cities and not just the national capital or a few state capitals. This makes it imperative that the role of external subsidy be minimised and plans developed that are locally sustainable.

Transportation "needs" cannot be considered an independent variable. These needs are greatly influenced by urban form and access policies. It has been shown that people may use different modes and technologies, public or personal transport, but the daily average motorised travel time varies relatively little. The travel time for all modes, including journeys by foot and by bicycle, differ little as everyone has the same time constraints – twenty-four hours with an 8–10-hour work-day. In low-density cities that are car dominated and have a high amount road space, car owners travel long distances relatively quickly, but have high travel times (like most cities in the United States and Australia). It is difficult to establish efficient public transport systems in such cities. Those who do not have cars cannot get to work easily and can thus be excluded from economic activities. On the other hand, those cities that have a dense settlement pattern tend to be more public transport-friendly and more citizens use non-motorised forms of transport (Tokyo, Singapore and Hong Kong).

These worldwide experiences tell us that access *needs* (as opposed to mobility patterns) remain similar across cities of different sizes and populations. Basically, during the peak period, every worker and student needs to get to work and school/college using one trip each. Since transport needs have to be worked out for the peak period, one can say that needs do not change per person over time. What may change are modal shares – and this can be influenced by policy. Since the travel budget remains the same for most people, the majority of trip distances differ not so much by the change in city size, but more by the change of mode. In India, the majority of trip distances are still less than 6–7 km.

All the old European and American cities that have extensive rail-based public transport developed in much the same way in the period 1900–1950, before large-scale private vehicle ownership. This resulted in the natural development of CBDs, which were fed by metro systems. Typically, cities such as Tokyo, New York, Paris and London provided exceedingly large CBDs. These cities have CBDs with more than 750,000 jobs. The massive CBD employment numbers and densities support a high degree of substitution by rail of automobile use that is not possible in smaller CBDs (because there is too little demand and they are too dispersed). The prerequisites for rail success are thus high residential population density and massive CBDs.

There are no cities in India where the CBD is expanding or increasing in density to provide 750,000 jobs. According to the 2001 Census of India, the main workers in India constitute about 30–35% of the population. Of these workers, a significant proportion (say about 30%) are self-employed, daily wage labour, hawkers, etc., who are not likely to use any form of motorised transport. Therefore, workers as the potential motorised transport users would amount to about 20% of the population. This means, that for a city to have a CBD to attract 750,000 workers by motorised means, it would need to have a population in excess of about 5 million, and the city should have only one major business district.

Since development in all Indian cities is taking place on the periphery, it is unlikely that any city India will develop a CBD of the size mentioned above. This is not surprising as all Indian cities have been planned as polynuclear cities. If this is so, no Indian city will be able to feed very high-capacity (>30,000 passengers per hour per direction [pphpd]) mass transit systems. This why the two metro systems built in India have not been able to attract sufficient numbers of passengers. The metro in Kolkata is carrying just 10% of the number of passengers that was projected when the project was sanctioned. The Delhi metro project was sanctioned on the basis of projections that it would carry 2.2 million passengers per day when Phase I was completed in December 2005. However, on completion, it has been able to attract only 0.4 million passengers a day (3–4% of all trips in Delhi) after an expenditure of Rs 120 billion on the infrastructure. This amounts to a subsidy of about Rs 30,000–40,000 per person per year on the cost of capital alone. The Delhi Metro Rail Corporation (DMRC) also claimed that, on completion of the project, there would be 2,500 buses less on the road and both pollution and road accidents would reduce significantly. In fact all three have increased, with the DMRC adding more feeder buses itself. Therefore, the project has not met any of its claimed benefits.

This is not surprising. It is the same story in other LMIC cities. Elevated or underground systems are justified on the basis of peak carrying capacities of 50,000 to 60,000 pphpd. However, like Delhi and Kolkata, the Bangkok subway is operating at peak loads of only 15,000–20,000 pphpd, the Kuala Lumpur metro at 10,000 pphpd and Mexico City, which has 201 km of metro rail and is the cheapest in the world, carries only 14% of trips. It is quite clear that cities that have developed after the 1950s and have multiple business districts will not be able to support very high-capacity metro systems either in terms of passenger loads or financial viability. Therefore, all metro projects planned in India will end up being white elephants at the expense of taxpayers' money!

International evidence suggests that, with modern communication systems, smart card ticketing, the use of the global positioning system (GPS), intelligent transport technologies and computer optimisation techniques, it has become possible to serve urban transport needs with modern bus rapid transit (BRT) very adequately. BRT systems need segregated lanes for buses, which reduce friction with other road users. These bus systems have the following advantages.

BRT systems can serve the needs of medium-sized cities all the way up to megacities with a dense network of routes going close to passengers' homes and destinations.

BRT systems can easily reach capacities of 20,000 pphpd. These capacities are very adequate for cities that have multiple business districts and medium-rise buildings. In such cities, higher-capacity systems will never run at peak capacity. However, the experiences of cities such as Bogota and São Paulo demonstrate conclusively that capacities up to 40,000 pphpd are feasible when catering for high-density CBDs.

BRT can be implemented at less than one-twentieth of the cost of building metro rail transit and light rail systems, typically Rs 5–10 crore per km.

BRT systems use existing rights-of-way on urban corridors and so the modifications involved do not disrupt the city significantly.

A major advantage of BRT is its flexibility in meeting changes in a city's development and changes in demand in quality and quantity. Expanded or new services can be introduced whenever needed. Unlike fixed track systems, BRT does not fix the city structure for ever.

When road systems are modified for BRT, it results in complete urban renewal as a part of the BRT project. This does not happen in the case of rail systems.

A BRT line can be constructed and opened in 18 months compared to 5–7 years for a metro line. This is a tremendous advantage, as efficient public transport becomes available before more people buy and get used to personal vehicles.

BRT serves the purpose of being a low-cost, medium-capacity public transport system that can be built quickly on the surface and in stages. Surface corridors need to have bus stops on the surface, promote street business development all along the corridor (not only at large metro stations) and provide eyes on the road (bus passengers and staff). This has the result of reducing crime and increasing socialisation, and thus increasing pedestrian and cycling trips. With segregated bus lanes and a provision for bicycling, safety improves, further improving the chances of reduction in motor vehicle use.

For growing cities of the twenty-first century, the debate is not between rail-based metro systems versus bus systems. The actual choice is between grade-separated systems and surface systems. You can use any technology on both. For all the considerations outlined above, the choice is clear – medium-capacity, dense network surface systems.

CONCLUSIONS

- Safety from crime and road traffic crashes is a prerequisite for sustainable urban transport options. This needs businesses and/or street vendors along all avenues.
- Cities must have mechanisms for low-income citizens needs to get articulated and heard.
- Cities must have narrower avenues with smaller blocks (say 500 m) to allow for comfortable and safe access and promotion
- City design seems to influence all behaviour.
- At grade public transport systems with a dense network and public information systems integrated with ITS-enabled taxi systems seems to be the only way forward.
- All health-related negative aspects of car transport need to be given much more exposure – obesity, reduction in socialisation, excess water consumption, etc.

REFERENCES

Bannister, D. 2005. *Unsustainable Transport: City Transport in the New Century.* Routledge, New York.

Government of India. 2001. *Census of India 2001: Provisional Population Totals.* Registrar General and Census Commissioner of India, Ministry of Home Affairs, New Delhi, India.

NEERI. *Ambient Air Quality Status for Ten Cities of India 2002. Report 12.* 2003. National Environmental Engineering Research Institute, Nagpur, India.

Ojha, A.K. 2002. *Road Accidents in Delhi 2001.* Delhi Traffic Police, New Delhi, India.

Planning Department. 2005. *Economic Survey of Delhi, 2003–2004.* Government of the National Capital Territory of Delhi, Delhi, India.

Saari, A., Lettenmeier, M., Pusenius, K. and Hakkarainen, E. 2007. Influence of vehicle type and road category on natural resource consumption in road transport. *Transportation Research Part D: Transport and Environment* 12, 23–32.

Stockholm County Council. 2003. *Munich-Stockholm – Comparison of the Two Regions' Planning Systems and Contents.* Regionplane-Och Trafikkontoret, Stockholm, Sweden.

Stockholm County Council. 2005. *Statistics on the Stockholm Region 2003/2004.* Regionplane-Och Trafikkontoret, Stockholm, Sweden.

Tegner, G. 2003. The Impact of Road Safety Interventions in Stockholm: Linking Road Demand, Accidents, Severity and Speed in a Model. Transek Consultants, Solna, Sweden.

The World Bank. 2006. *2005 World Development Indicators – Environment: Air pollution.* The World Bank, Washington D.C.

WBCSD. 2004. *Mobility 2030: Meeting the Challenges to Sustainability.* World Business Council for Sustainable Development, Geneva, Switzerland.

Index